Innovation and the Multinational Firm

Also by Alessandra Perri

MANAGING INNOVATION IN EMERGING ECONOMIES: Organizational Arrangements and Resources of Foreign MNEs in the Chinese Pharmaceutical Industry (*with Scalera V.G. and Mudambi R., 2015*)

KNOWLEDGE SPILLOVERS FROM FDI: A Critical Review from the International Business Perspective (*with Peruffo E., 2015*)

A LONGITUDINAL STUDY OF MNE INNOVATION: The Goodyear Tire and Rubber Company (*with Scalera V.G., Mukherjee D. and Mudambi R., 2014*)

KNOWLEDGE OUTFLOWS FROM FOREIGN SUBSIDIARIES AND THE TENSION BETWEEN KNOWLEDGE CREATION AND KNOWLEDGE PROTECTION: Evidence from the Semiconductor Industry (*with Andersson U., 2014*)

BALANCING THE TRADE-OFF BETWEEN LEARNING PROSPECTS AND SPILLOVER RISKS: MNCs' Subsidiaries Linkage Patterns in Developed Countries (*with Andersson U., Nell P.C. and Santangelo G., 2013*) palgrave▸pivot

Innovation and the Multinational Firm: Perspectives on Foreign Subsidiaries and Host Locations

Alessandra Perri
Universita' Ca' Foscari Venezia, Italy

palgrave
macmillan

© Alessandra Perri 2015

All rights reserved. No reproduction, copy or transmission of this publication may be made without written permission.

No portion of this publication may be reproduced, copied or transmitted save with written permission or in accordance with the provisions of the Copyright, Designs and Patents Act 1988, or under the terms of any licence permitting limited copying issued by the Copyright Licensing Agency, Saffron House, 6–10 Kirby Street, London EC1N 8TS.

Any person who does any unauthorized act in relation to this publication may be liable to criminal prosecution and civil claims for damages.

The author has asserted her right to be identified as the author of this work in accordance with the Copyright, Designs and Patents Act 1988.

First published 2015 by
PALGRAVE MACMILLAN

Palgrave Macmillan in the UK is an imprint of Macmillan Publishers Limited, registered in England, company number 785998, of Houndmills, Basingstoke, Hampshire RG21 6XS.

Palgrave Macmillan in the US is a division of St Martin's Press LLC, 175 Fifth Avenue, New York, NY 10010.

Palgrave Macmillan is the global academic imprint of the above companies and has companies and representatives throughout the world.

Palgrave® and Macmillan® are registered trademarks in the United States, the United Kingdom, Europe and other countries.

ISBN: 978-1-137-55545-8   EPUB
ISBN: 978-1-137-55544-1   PDF
ISBN: 978-1-137-55543-4   Hardback

A catalogue record for this book is available from the British Library.

A catalog record for this book is available from the Library of Congress.

www.palgrave.com/pivot

DOI: 10.1057/9781137555441

*To mum Franca and dad Gabriele,
with unconditional love and gratitude*

Contents

List of Illustrations   viii

Acknowledgments   ix

List of Abbreviations   xii

Introduction   1

Part I Managing Innovation across Geographical Space: An Overview

1   Innovation in Multinational Firms and the Role of Geography   9
    1.1   Innovation and the multinational firm   10
    1.2   Models and processes of innovation in multinational firms   12
    1.3   R&D internationalization and the role of geography   15
          1.3.1   Tacit knowledge, proximity and co-location   16
    1.4   Is R&D really international?   22

2   Managing Innovation in Multinational Firms   34
    2.1   Forces behind R&D internationalization   35
    2.2   Types of innovation-driven foreign direct investment   40
    2.3   Location choices of R&D facilities   45
    2.4   Organizational challenges in multinational innovation 52
          2.4.1   Orchestrating geographically dispersed innovation activities   54

Part II A Multilevel Approach to the Study of Geographically Dispersed Innovation in Multinational Firms

3 Perspectives on Subsidiaries — 62
- 3.1 The evolution of subsidiary-level research — 63
- 3.2 Changing roles: from passive to active subsidiaries — 65
- 3.3 Internal and external drivers of subsidiary evolution — 72
  - 3.3.1 The influence of the internal network — 72
  - 3.3.2 The influence of the external network — 75
- 3.4 An innovation management perspective — 77
  - 3.4.1 Knowledge creation — 80
  - 3.4.2 Knowledge protection — 83

4 Perspectives on Host Locations — 86
- 4.1 The relevance of locations for MNC innovation — 87
- 4.2 Geographical systems of innovation — 91
- 4.3 Agglomeration, clusters, cities — 94
- 4.4 Place and space: locational features for MNC innovation — 101

5 Integrating Perspectives — 105
- 5.1 Integrating international business, innovation and economic geography perspectives — 106
- 5.2 The case of FDI spillovers to host locations — 110
- 5.3 Antecedents of FDI knowledge spillovers — 113
  - 5.3.1 Macro-level perspectives — 113
  - 5.3.2 Meso-level perspectives — 115
  - 5.3.3 Micro-level perspectives — 117
- 5.4 Implications for the study of FDI spillovers to host locations — 121

6 Concluding Remarks and New Research Directions — 124
- 6.1 Concluding observations — 125
- 6.2 Potential research opportunities — 130

Bibliography — 133

Index — 157

List of Illustrations

Figures

2.1	Type of foreign innovative units by geographical scope and level of technological competence	44
3.1	The relationships between subsidiary-level and location-level conditions and objectives in subsidiary innovation strategy	79
5.1	A multilevel approach to the analysis of geographically distributed innovation in MNCs	109
5.2	Subsidiary strategy and conduct and local knowledge spillovers	122

Tables

1.1	Business sector R&D expenditure by affiliates abroad as a percentage of domestic R&D	25
1.2	R&D expenditures of foreign firms as a percentage of total business R&D expenditure, and growth rate	26
1.3	Share of patents invented abroad, and growth rate	28
1.4	Share of patents owned by foreign residents, and growth rate	30

Acknowledgments

The ideas developed in this book result from a long process, whose maturation took place during my stay at the Department of Strategic Management of the Fox School of Business at Temple University. In the months spent at Philadelphia, I have had the pleasure to work and interact with Ram Mudambi. Since I met him, Ram has been a great source of knowledge and inspiration. The research scope and theoretical approach proposed in this volume have been strongly influenced by the research agenda of the iBEGIN (International Business, Economic Geography and Innovation) research group of which I am part, and that Ram coordinates with great enthusiasm. The conversations and long research meetings that I have had with Ram and with other iBEGIN members during my stay at Temple have represented the lifeblood of this book. But Ram offers to young scholars much more than his research experience: his helpfulness, optimism, curiosity and generosity make him an ideal mentor, but most importantly an invaluable friend.

Along with Ram, Philadelphia endowed me with another great colleague and friend, Vittoria Scalera. Vittoria is the person with whom I shared many ideas and conversations on issues referring to emerging market locations. She has been the perfect companion for both the hard-work and the fun that characterized my stay at Philly. I am grateful to have met her, and I hope that our collaboration and friendship will always be as pleasant and productive as it has been so far.

Many thanks also go to the other brilliant members of the iBEGIN research group, and particularly to Marcelo Cano Kollman, T.J. Hannigan, Arheum Lee and Eunkyung Park.

Much before Philadelphia, my research ideas have been inspired by another great scholar: Ulf Andersson. Ulf has been a friend even before becoming a coauthor. His contribution to my view of subsidiaries and local business networks has been key for the development of many of the ideas developed in this book. Hence, I am very grateful to have met him during my PhD, when his guidance and support helped to overcome many difficulties.

I am also very thankful to have the opportunity to interact with Grazia Santangelo, whose dedication to research has represented a spur to improve my own research skills since the very first moment I met her. Grazia's suggestions have contributed to develop many of my ideas and works, and have supported me also in the writing on this book. As a female researcher, I am very happy to have a reference model like her.

I am also indebted to Larissa Rabbiosi who has been so kind and generous to share with me her comments on the very first project of this book. Since I first met her during my stay at Copenhagen Business School, I have always appreciated her sincere willingness to help and give genuinely valuable suggestions.

As the roots of this book go back to my years as a PhD student, I wish to thank Enzo Peruffo, a friend and colleague with whom I have had long conversations on issues included in this book, and particularly on the topic of FDI spillovers. Enzo is the type of friend you really want to have around during your PhD, and I am blessed to have the opportunity to rely on his sincere support.

I also wish to thank my PhD supervisors, Matteo Caroli and Raffaele Oriani, for the guidance they offered me in the development of my dissertation, which laid the basis for many ideas developed in this book.

More recently, my research on multinationals and emerging-country locations has benefited from the much appreciated comments and suggestions of Tiziano Vescovi, to whom goes my sincere gratitude.

I am also grateful to Claudio Giachetti and Francesco Zirpoli, for their valuable suggestions in the stage of design of this book.

I have had the opportunity to work with very professional, but most importantly extremely helpful people at Palgrave Macmillan. My special thanks go to Liz Barlow and Maddie Holder.

Behind the writing of a book, there are moments of confusion, hesitation and even frustration. But if you can rely on the openhearted support of good friends, these moments vanish. In this regard, I am strongly indebted to my colleagues Anna Moretti, Elisa Cavezzali, Francesca Checchinato and Gloria Gardenal.

My family in general and my parents in particular are the persons to whom I am mostly indebted, as they have always supported my choices, and ensured to make my path as smooth as they could, and even more. They have always been enormously flexible, open-minded and sympathetic to understand the admittedly weird habits of a young academic.

Finally, my thanks go to Salvatore, to whom the writing of this book has stolen time and attention.

The responsibility for any mistake or omission remains mine.

List of Abbreviations

EG	Economic Geography
FDI	Foreign Direct Investment
GDP	Gross Domestic Product
GERD	Gross Expenditure in Research and Development
IB	International Business
IP	Intellectual Property
IPR	Intellectual Property Right
IS	Innovation Studies
MNC	Multinational Corporation
NIS	National Innovation System
R&D	Research and Development
RIS	Regional Innovation System
SIS	Sectoral Innovation Systems
TIS	Technological Innovation System
US	United States

Introduction

Abstract: *This introduction proposes a brief synopsis of the main topics and arguments covered in this book. The book is divided into two parts. The first part offers an overview on the nature and the dynamics of innovation in multinational firms, analyzes recent trends in R&D internationalization and discusses the wide-ranging implications arising from the management of innovation across geographical space (Chapters 1–2). The second part emphasizes the increasingly crucial role of foreign subsidiaries and host locations, and systematizes relevant theoretical perspectives and levels of analysis to develop an analytical framework for the study of geographically distributed innovation in multinational firms (Chapters 3–5). Chapter 6 concludes and offers directions for future research.*

Keywords: chapters' overview; geography; innovation; multinational firms

Perri, Alessandra. Innovation and the Multinational Firm: Perspectives on Foreign Subsidiaries and Host Locations. Basingstoke: Palgrave Macmillan, 2015. DOI: 10.1057/9781137555441.0005.

Innovation is critical to generate value, and multinational corporations (MNCs) have long established themselves as main actors in the process of advancement of the technology frontier. Traditional literature on the management of innovation in multinational contexts has investigated how MNCs organize their knowledge creation activities and how these choices affect their innovation performance. However, there is by now widespread recognition that MNC corporate-level strategies and activities are not the only relevant phenomena that have to be studied, if we aim at gaining a fuller understanding of how innovation is created and managed within MNCs. International business literature has progressively highlighted the critical role that the network of geographically distributed subsidiary units plays in the development of value-generating activities. Similarly, both international business and economic geography scholars have recently suggested that in order to fully appreciate the interplay between MNC strategy and its spatial environment, a more nuanced view of locations is needed.

This book aims at offering up-to-date insights on how innovation is managed within MNCs[1], by paying particular attention to these two aspects, namely (1) the role of MNC subsidiaries, as vital actors in the MNC innovation process and in their own local innovation networks, and (2) the characteristics of host locations that, at different geographical scales, influence and are in turn influenced by the innovative activities of MNC subsidiaries. In doing so, the book assumes a novel theoretical angle, inspired by the blend of three literatures that are critical to examine the complex dynamics of innovation in MNCs, namely international business (IB), innovation studies (IS) and economic geography (EG). Therefore, combining existing theoretical perspectives whose integration may be conducive to broader understandings, it intends to emphasize the fundamental importance of bringing the analysis of geographically distributed innovation in MNCs closer to the level in which it actually happens, that is, the level of MNC subsidiaries and host locations, and of the complex interactions between the two.

For this purpose, the book focuses on subsidiaries as strategizing actors of both the internal (MNC) and external (host location) networks to which they participate, thus highlighting the opportunities and challenges of conducting innovative activities in foreign locations, and the wide-ranging incentives to which they are exposed and that inform their decision making and behavior. Likewise, leveraging recent insights pervading both IB and EG research, the book highlights the importance of treating locations as "spaces", thus accounting for the diverse set of locational features operating at different geographical levels of analysis (national systems of innovation, clusters, cities) that are key to subsidiaries' innovation processes. Moreover, answering to recent calls for a more fine-grained account of both spatial heterogeneity at the subnational level and spatial discontinuities at the national level, it explains that, in order to better inform the analysis of subsidiary innovation, it is pivotal to distinguish between "border effects", widely investigated by IB scholars, and "distance effects", emphasized by regional science and EG literature streams.

The book is divided into two parts. The first part includes Chapters 1 and 2, and is intended to offer a general perspective on the nature and the dynamics of innovation in MNCs, to analyze the most recent trends in the internationalization of research and development (R&D) and to discuss the strategic and organizational implications arising from the management of innovation across borders. The second part comprises Chapters 3, 4 and 5, and aims at emphasizing the role of MNC subsidiaries and host locations in innovation processes, thereby reconstructing the complex interplay among theoretical perspectives and levels of analysis to ultimately develop an analytical framework for the study of geographically dispersed innovation in MNCs.

Chapter 1 provides a broad view on the role and the characteristics of innovation within MNCs. To start and put the argument into context, it discusses the link between innovation and MNCs. Following established literature, innovation is presented as the rationale for the very existence of MNCs and the key factor for creating value and strengthening MNCs' competitive advantage, while MNCs are portrayed as organizations relying on two networks that are critical for knowledge sourcing and knowledge creation: the internal network, composed of the headquarters and the geographically distributed MNC subsidiaries, and the external network, made up of the local organizations with which subsidiaries develop linkages to better manage the foreign environment. With the aim of emphasizing the critical role of geography for the management of innovation in MNCs, this chapter works with diverse theoretical perspectives that enable to understand how this role has changed over time. While some scholars have put forward the idea of a "flat world", in which the progress in communication and transportation technologies and infrastructure makes distance fundamentally irrelevant, other scholars have supported the view that proximity is still critical for the effectiveness of particular activities. In the context of MNC innovation, geography and proximity are presented as critical factors for the management of both the external and the internal MNC networks. On one hand, because knowledge is sticky and tends to follow patterns of specialization in specific regions, firms may effectively tap into geographically distant bodies of knowledge and expertise only by internationalizing their R&D. On the other hand, as MNCs internationalize their R&D, activities that are critical to the effective management of innovation within the MNC organizational boundaries, such as control, communication and knowledge sharing, become increasingly complex. Related to the long-lasting trade-off between innovation concentration and geographical dispersal (analyzed in Chapter 2) is the debate on the extent to which R&D has actually become *international* over the years. Confronting opposite viewpoints in the literature, the chapter brings to the reader fresh information and interpretations on R&D globalization trends.

Chapter 2 draws on existing literature to describe the most important corporate-level decisions and organizational issues associated with the internationalization of innovative activities by MNCs. Whether and how to establish a knowledge-intensive subsidiary abroad is a critical choice that is demanded to headquarters. Moreover, because knowledge is sticky and does not flow easily even within the organizational boundaries, headquarters are required to coordinate and encourage inter-unit knowledge transfer. The objective of this chapter is precisely to elucidate the role of MNC headquarters in the management of (potentially distributed) innovative activities. Geography and proximity return at the center of the discussion here because, as introduced in Chapter 1, both R&D concentration and its geographical distribution lead to advantages and disadvantages that have been widely investigated by existing literature. Hence, centripetal and centrifugal forces that influence the MNC geographical organization of R&D activities are discussed. The chapter also focuses on the most relevant external and internal drivers that, over the years, have been found to encourage overseas innovative investment. Accordingly, the different motivations for establishing value-generating activities abroad, and the resulting empirical R&D subsidiary types are reviewed, explaining the rationale behind them as well as the challenges and opportunities they encompass. The chapter also describes the most critical dilemmas faced by headquarters: (1) when designing the knowledge-intensive foreign direct investment (FDI), and (2) during the subsequent management of dispersed innovative units. As far as the R&D foreign investment design is concerned, special attention is paid to the issue of location choices, reviewing and systematizing various angles and theoretical perspectives that have contributed to elucidate this critical element of MNC strategy-making. Following the analysis of R&D foreign investment design stage, the chapter considers the organizational challenges that may arise after the establishment of foreign R&D subsidiaries. Specifically, problems stemming from the management of control, communication and coordination are discussed in the light of IB and IS perspectives. Overall, this chapter examines the configuration and management of international innovative activities as a complex process that exposes MNC headquarters to a number of critical choices and trade-offs, which need to be addressed in consideration of both internal and external constraints and opportunities.

Chapter 3 aims at emphasizing the implications of the insights offered by the growing body of studies on the role of subsidiaries in MNC innovation. Traditional literature depicts subsidiaries as passive actors of both the MNC internal organization and the external network of local organizations, whose role is limited to the mere implementation of headquarters' decisions. However, starting from the studies on subsidiary mandates and centers of excellence, scholars have increasingly reported evidence on subsidiaries' evolution toward a higher strategic sophistication. Accordingly, IB literature has called for a more inclusive analysis of MNC strategic decision process, covering not only the level of headquarters but also the subsidiary level. This chapter goes in this direction by consolidating existing literature on the changing role of subsidiaries and by providing an interpretative framework for the analysis of the driving forces of their local technological behavior. Factually, the chapter first organizes theoretical perspectives and empirical findings on the evolution of subsidiary roles and capabilities to introduce the distinction between competence-exploiting and competence-creating subsidiaries. The active role of subsidiaries is contextualized within the two networks in which they operate, namely their internal MNC network and their external local business network. After reviewing these insights, the chapter uses IS literature to offer an interpretative framework for the analysis of subsidiary innovation, discussing the subsidiary's most critical knowledge imperatives such as knowledge creation and knowledge protection. Consistent with recent research, this framework leads to identify subsidiaries as strategizing actors of their internal and external networks, and sheds fresh light on subsidiary-level heterogeneity. Accounting for these strategizing activities in subsidiary innovation management is critical to fully understand innovation in MNCs. Headquarters need to pay increasing attention to the potential effects of subsidiaries' innovation practices, and scholars may find additional research opportunities by exploring subsidiaries' strategizing activities in innovation-related matters further.

Chapter 4 intends to demonstrate the necessary link between the literature on IB and IS on one hand, and the EG perspective on the other hand, embracing recent views suggesting that EG insights are essential to understand MNCs' geographical behavior. The notion of location in IB literature has been mainly conceptualized and empirically identified with the national scale. This has happened despite the fact that MNCs often establish multiple units in different geographical areas of the same country, which suggests that sub-national heterogeneity matters, beside cross-national heterogeneity. Sub-national heterogeneity is likely to play a role particularly when innovative activities are concerned, given that knowledge flows are geographically bounded. Therefore, with no claim of exhaustiveness, the chapter works with selected relevant sub-national scales and related theoretical perspectives that may help explain MNCs' attraction to and interaction with specific external environments. Moreover, because EG scholars have recently suggested that conventional models of knowledge diffusion within localized territorial units are plagued by limitations that do not allow to properly understand the wide-ranging potential advantages and disadvantages of co-location, the chapter also discusses different sources of locational heterogeneity that may explain uneven patterns of knowledge diffusion and creation within particular territorial units. In doing so, it covers issues such as the technological profile of agglomerated actors, the governance of local transactions and the structural properties of locational networks. Particular emphasis is given to agents and mechanisms that foster the openness and porousness of locations, as scholars from different disciplines increasingly recognize the pivotal role of extra-local relationships in the renewal and vitalization of local knowledge bases. On the whole, the chapter proposes that the dynamic interaction between the variety of locational features of different spatial configurations and the MNC's strategy and conduct is an essential building block for the analysis of MNC innovation.

Chapter 5 integrates the arguments developed in Chapters 3 and 4 to propose an analytical framework for the study of geographically distributed innovation in MNCs. It is argued here that, as both the increasingly active role of MNC subsidiaries and the critical importance of sub-national geographical scale gain wide recognition in IB and EG literature, the innovation processes that occur in geographically dispersed locations of MNCs can only be analyzed by lowering the level of the analysis to where the interaction between subsidiaries and sub-national host locations actually occurs. This implies focusing on wide-ranging sources of subsidiary-level and location-level heterogeneity. Simultaneously, it also requires accounting for the higher-level systems in which both subsidiaries and sub-national locations are embedded, whose influence is critical to fully appreciate the nature and the rationale of specific patterns of subsidiary-location interactions. To offer an example of how this analytical framework can be applied to issues relating to the geographical dispersal of innovation in MNCs, the second part of the chapter is devoted to the analysis of a critical research problem in IB literature, that is, FDI knowledge spillovers. This exercise suggests that a comprehensive account of all the relevant levels of analysis can strongly modify the predictions of more conventional IB models, and it shows that just like locations and locational features are important for MNC and subsidiary innovation, the latter may be critical for the locations' evolution and performance.

Chapter 6 concludes and discusses the analytical framework proposed in the book, offering insights into how the arguments developed in the previous chapters can be used to develop new research directions at the intersection between IB, IS and EG.

Note

1 It is useful to anticipate here that while the book often refers specifically to R&D activities, it intends to take a broader perspective on innovative activities, intended as the wide range of activities through which new knowledge is created.

Part I
Managing Innovation across Geographical Space: An Overview

1
Innovation in Multinational Firms and the Role of Geography

Abstract: *Chapter 1 discusses the interrelations between innovation and the MNC, reviews selected models of innovation in MNCs and analyzes the role geography plays in shaping the patterns of knowledge diffusion. Building on previous research, it suggests that the development of technological capabilities and the international expansion are mutual-reinforcing strategic paths. Moreover, despite recent views proposing that geography is becoming increasingly irrelevant, it contends that geographical proximity still plays a powerful role in determining economic activities and, more specifically, in influencing MNCs' management of internal and external innovation networks. Using OECD data, it concludes by offering some insights into recent trends of R&D internationalization.*

Keywords: co-location; proximity; R&D internationalization; tacit knowledge

Perri, Alessandra. *Innovation and the Multinational Firm: Perspectives on Foreign Subsidiaries and Host Locations.* Basingstoke: Palgrave Macmillan, 2015. DOI: 10.1057/9781137555441.0007.

1.1 Innovation and the multinational firm

In the current competitive landscape, the ability to innovate constantly despite global economic downturns and changing industry dynamics is a crucial driver of value creation (Scalera et al., 2014). Multinational corporations (MNCs) have been very effective in the development of such capability, as revealed by several insightful facts:

- MNCs are the dominant spenders in research and development (R&D) activities; for instance, in 2011, many of the top 20 global R&D spenders, all of which are MNCs, spent in R&D more than the total gross domestic expenditure on R&D (GERD) of countries such as Hungary, Ireland, Portugal, the Czech Republic and Norway;[1]
- in 2014, the top 20 global R&D spenders, all of which are MNCs, accounted for approximately 25% of the overall R&D spending of the top 1000 corporate R&D spenders, which in turn explains about the 40% of the world's R&D spending (Jaruzelski et al., 2014);
- in 2014, the top 25 US patent assignees were MNCs (IFI Claims, 2015);
- from 2005 to 2014, the 13 corporate spenders[2] that have structurally maintained a position among the top 20 R&D spenders are MNCs (Jaruzelski et al., 2014).

As these data show, MNCs can be considered as the prime developers of technological capacity (Dunning, 1994). Hence, it is important to investigate how R&D and innovation activities are managed within these organizations.

International business (IB) literature has long highlighted the existence of a strong association between innovation processes and the MNC. On one hand, R&D is among the possible sources of ownership advantages that ease firms' international development (Caves, 1982). On the other hand, a company's technological capabilities can be intended as being highly firm-specific, thus requiring an internal organization of the underlying international transactions which constitutes a major driver of firms' foreign expansion (Buckley and Casson, 1976). In addition, through their international presence, MNCs may gain access to foreign knowledge thus strengthening their technological advantages (Dunning, 1993; Kuemmerle, 1997).

Although these arguments should not advice toward the existence of an unambiguous causal link between technological competence and firm internationalization, whose direction would be very difficult to establish, it is undeniable that within the wide-ranging processes of technological accumulation that drive organizations to constantly build and renew their resources and capabilities to face global competitive pressures, the investment in technological competencies and the international expansion are two crucial, and often mutual-reinforcing, strategic paths (Cantwell, 1995a).

In other words, as MNCs develop the capability to accumulate, enrich and exploit technological knowledge, their geographic scope tends to expand. Concurrently, the exposure to foreign knowledge sources enables MNCs to be continuously responsive to state-of-the-art technologies originating anywhere in the world, thereby presiding the innovation frontier.

The MNC technological innovation can be conceptualized as the outcome of a complex process involving technological accumulation and collective learning. Within this process, R&D plays several critical roles, which span from the support to current production needs to the pursuit of long-term objectives of new knowledge creation. In the first case, R&D usually operates in close connection with specific production facilities to the aim of responding to local conditions. Indeed, to convert knowledge into products and services dedicated to foreign markets, MNCs need to carry out processes of adaptation. In the second case, R&D is used to source external knowledge and monitor the emergence of technological opportunities, whose commercial exploitation is only potential.

Traditionally, the need to adapt products and processes to conditions in the foreign markets has represented the spark that ignited R&D internationalization. While the majority of MNCs' value chain activities have rapidly experienced an increasing geographical dispersal aimed at capturing opportunities arising from foreign countries' comparative advantages, MNCs have long tried to keep their R&D operations within the home-country boundaries. Yet, in the last decades, it became increasingly evident that proximity to manufacturing plants is essential to effectively target regional markets. The resulting R&D internationalization process has gradually evolved to incorporate other R&D activities beyond adaptation, such as scouting of emerging technologies and foreign knowledge acquisition.

This phenomenon poses several managerial and organizational challenges, but simultaneously encompasses a number of important benefits that have attracted scholars' attention. The wider picture in which these dynamics occur is one that witnesses the evolution of MNCs into increasingly interactive and internally differentiated networks aimed at the creation of new capabilities, in which specialized activities are performed in particular geographical sites characterized by the endowment with location-specific knowledge (Zanfei, 2000).

The gradual establishment of this geographically dispersed system for corporate development underscores the importance of exploring not only the role of foreign locations, but also the actors that are required to manage MNCs' innovative activities outside the home-country, that is, foreign subsidiaries.

1.2 Models and processes of innovation in multinational firms

To understand the contemporary organizational and strategic issues that characterize the management of innovation in MNCs, it is important to recall the origins and the evolution of this phenomenon, as well as the most important theoretical approaches that have been developed for its interpretation.

To this aim, it is useful to start from the contribution of Vernon who, in 1966, proposed a theory of the product cycle model, thereby offering an interpretation of international technology flows. According to the author, firms in advanced, industrialized countries are likely to be more sensitive to the detection of opportunities for novel products. Therefore, advanced countries not only spur the innovation process but also represent the preferred setting for the actual development of innovation, because they offer extensive pools of skilled human resources, interaction opportunities with prospective customers and smooth communication among innovative actors and among these and production facilities. As a demand for products embodying the new technology emerges in foreign countries, the innovation may be transferred into other locations through export modes, technology-based market transactions such as licensing and, ultimately, through the establishment of overseas subsidiaries.

It is worth mentioning here that the original product cycle model is based on the idea that innovation is a demand-driven, linear process.

Indeed, the underlying assumption of this approach is that innovations are stimulated by particular conditions emerging *from the marketplace* (Vernon, 1979). This is the reason why Vernon considered the US market as the most convenient innovation source for indigenous MNCs, which could leverage the sophisticated needs of a high-income market and the critical expertise of advanced downstream manufacturing facilities. From an economic geography (EG) perspective, Vernon's model has the merit of highlighting the spatial consequences, in terms of shifts in firms' operations, arising from an industry's development cycle, thereby denoting the existence of a sort of hierarchical structure in the geographical location dynamics of international production (Iammarino and McCann, 2013).

Despite his pioneering focus on MNCs' activities spatial movements, which inspired scholars in both regional and urban economics, subsequent IB studies building on Vernon's model did not explore the space-related insights envisaged by his works (Iammarino and McCann, 2013). For instance, Kindleberger (1969) and Stopford and Wells (1972) developed theories that are close to the product cycle model but mainly focus on organizational issues; in their view, subsidiaries are mere implementers of the headquarters' strategic choices, including those involving the firm technological activities.

While the explanatory power of these models was initially quite strong, several conditions have gradually changed, driving to reconsider the validity of the underlying approach. Specifically, it became manifest that innovation processes are not confined to firms' home-countries anymore, as MNCs knowledge-based activities increasingly draw upon globally distributed networks of subsidiaries (Vernon, 1979).

Several empirical studies confirmed the national boundary spanning process that firm innovation activities were undergoing. In 1990, Ghoshal and Bartlett propose a classification of the different innovation tasks performed by MNCs' subsidiaries in three main categories: creation, adoption and diffusion. The "creation" task refers to the autonomous development of new products and processes to respond to local contingencies, and hints at the organization's responsiveness, intended as the ability to use the available resources to exploit opportunities arising from diverse locations. The "adoption" task covers the range of activities that subsidiaries are required to execute to embrace and implement innovations generated by the headquarters or by other sister units, thus being closely related to the pursuit of an integrated global strategy at the firm level. Finally, the "diffusion" task relates to those situations in which subsidiaries are requested to share their locally created innovations with the parent-company or with other subsidiaries, thus allowing for the exploitation of the individual subsidiary's knowledge in a wider range of geographically distributed MNC nodes.

Clearly, the introduction of "creation" tasks significantly expands the functions that were traditionally ascribed to foreign subsidiaries in their roles of replicators of the parent-company's activities abroad. In particular, the recognition of the variety of innovative activities that subsidiaries may be requested to undertake hints at the diversity of the routes that innovation may follow within MNCs' boundaries. In this regard, Bartlett and Ghoshal (1990) propose four different processes of MNC innovation that, however, should not be considered as mutually exclusive, as MNCs should be able to implement them simultaneously to pursue different projects. All four processes encompass both advantages and drawbacks, which need to be evaluated in combination with the specificities of the firm and of the innovation project itself.

The two most traditional processes have been labeled as "center-for-global" and "local-for-local". The "center-for-global" process identifies in the corporate headquarters the most critical actor of innovative activities. According to this view, it is the MNC central R&D laboratory to be provided with the critical resources and capabilities to develop innovation that will be subsequently leveraged worldwide through a network of geographically distributed subsidiaries. Despite the advantages of centralization, such process and the resulting innovation structure entail the risk of disregarding local market needs and encountering resistance on the part of subsidiaries, which may be reluctant to act as passive implementers of centrally imposed technologies.

In the "local-for-local" process, each subsidiary engages in the development of its own technological know-how, which enables them to create tailored innovations that fit with the needs and conditions in the host market. A potential drawback of this model lies in the inefficient use of MNC innovation resources, as different subsidiaries could develop their own answer to the same problem, or bear unnecessary differentiation costs across countries.

Along with these established innovation paths, empirical evidence has contributed to emphasize the existence of a "local-for-global" process, in which R&D and technological capabilities in specific subsidiary locations may lead to innovations endowed with worldwide competitive potential, which could therefore be used in the global market. Admittedly, the transfer and the adoption of local innovations by other local units could be very complex tasks, due to the well-known "not invented here" syndrome.

Finally, "global-for-global" innovation processes encompass the contribution and the effort of different R&D units to the aim of addressing a global problem. Despite its great potential in exploiting scope economies and leveraging worldwide learning, the feasibility and effectiveness of this type of innovation process is undermined by the high coordination effort required to connect different organizational components and point them toward the same, complex goal.

Along with Bartlett and Ghoshal's (1990) analysis of the diverse processes through which innovation can be developed in MNCs, Håkanson (1990) proposes a dynamic view of MNCs' organization of R&D activities and innovation across space. He suggests a three-stage model depicting the MNC's evolution from a centralized hub structure, where R&D is performed at the organization core and is supported by a network of small overseas R&D units dedicated to the fulfillment of adaptation needs and to the provision of technical assistance, toward a decentralized federation in which foreign research affiliates perform a wider range of tasks including more creative activities, which ultimately leads MNCs to operate as an integrated network where the headquarters is not the only developer and sender of new technologies, as the worldwide distributed subsidiaries are equally able to contribute knowledge to other parts of the network thanks to advanced internal communication and coordination systems.

1.3 R&D internationalization and the role of geography

As the foregoing discussion suggests, the predictions put forward by the product cycle model and by similar theories hinting at a hierarchical and unidirectional transfer of newly created technological knowledge from the MNC center to the periphery became increasingly less adherent to the reality of industrial R&D, which along the years experienced an internationalization process, triggered by the geographical choices of MNCs.

As highlighted by Cantwell (1995b), the fading relevance of Vernon's hypotheses depends on the empirical setting on which his observations were based, that is, the United States (US), where in the 60s local MNCs had a lower degree of foreign R&D compared to companies in other countries. Similarly, the beliefs regarding the demand-driven nature of innovation, which laid the basis for the product cycle model, gradually turned out to offer only a partial view of the innovation process (Mowery and Rosenberg, 1979), which in reality is propelled by firm-specific learning and by its interactions not only with the specificities of the market demand, but also with new scientific and technological knowledge (Cantwell, 1995b).

Consistent with the idea that both demand- and supply-side factors are critical to explain how firms' innovation activities evolve over time and space, value creation in MNCs increasingly relies on knowledge developed abroad, either through internal R&D processes conducted by foreign subsidiaries or by means of network relationships with external partners. Zanfei (2000) has conceptualized this phenomenon by acknowledging the shift of the MNC's organization toward a "double network" governance mode, emerging from the gradual diffusion of two critical dynamics.

A first dynamic has an internal nature, and it lies in the increasingly strong interrelation among the MNCs' units regarding the development and utilization of new knowledge. This interrelation enables to overcome the rigid flow of centrally created knowledge from the headquarters to the foreign subsidiaries and allows subsidiaries to participate to more creative activities, thus contributing to the knowledge-generation process of the entire organization.

A second dynamic has an external nature, and it refers to the emergence of a significant web of relationships that link the internal nodes of the MNC organization to agents located outside the firm boundaries, allowing both core and decentralized units to gain access to local sources of knowledge, expertise and information that are critical to feed the internal innovation process.

As we acknowledge that MNCs establish R&D subsidiaries abroad not only to adapt their products and processes to the local market but also for supply-driven motivations of foreign knowledge acquisition, geography and proximity emerge as main factors in the management of MNC innovation and of the firm external and internal knowledge networks.

1.3.1 Tacit knowledge, proximity and co-location

The reasons why geography and proximity matter in this realm are of course related to the nature of knowledge, to the way it evolves in time and space, and to the distinction between its tacit and codified components. This distinction has gained considerable relevance thanks to the works of prominent scholars such as Polanyi (1966), Nelson and Winter (1982), Nonaka and Takeuchi (1995). It was in fact Polanyi (1966) who captured the intrinsic character of tacit knowledge as he affirmed that people know more than they can express with words. Although it is neither possible nor appropriate to trace a clear borderline between tacit and codified knowledge, the former can be defined as knowledge that is not (easily) articulable, as opposed to knowledge that is more formalized and explicit (Nelson and Winter, 1982).

Literature has abundantly showed that the transfer of tacit knowledge requires processes of learning-through-interacting. Knowledge that is not codified can be effectively transmitted only through experience and practice. Moreover, knowledge exchange requires the sender and the receiver to share a common social context, and since the constituting factors of such a context tend to be defined at a local level, tacit knowledge is understood to be spatially sticky (Gertler, 2003). In other words, because social interaction is greatly facilitated by geographical proximity, the strong social component underlying the exchange of tacit knowledge suggests that the latter is difficult to transfer from a distance.

Several empirical studies demonstrate that technological knowledge is mainly tacit (see Collins, 1992, 2001 among others), thereby suggesting that tacit knowledge is a major determinant of the geography of technological innovation and, in turn, of value creation through technological innovation. In fact, prominent economic geographers such as Maskell and Malmberg (1999) have effectively explained that in an increasingly globalized economy, where *codified* technology is easier to obtain, the *tacit* component of knowledge assumes a fundamental role in nourishing firms' competitive position. As many skills and assets that were previously localized become available almost everywhere, the opportunity to access them does not grant any advantage. Conversely, tacit knowledge that is embedded in specific practices is still difficult to trade and transfer, despite globalization. Because it is not "ubiquified", thereby being accessible to only a handful of organizations, its vital role in ensuring distinctive competitive positions increases, and becomes more and more important as markets keep on internationalizing.

In contrast to this view, some researchers have recently promoted the idea of the "death of distance" (Cairncross, 1997; Friedman, 2005), according to which the progress in information, communication and travel technologies enables knowledge to flow easily around the world, leaving no role for geography in the dynamics of its diffusion process and, consequently, in the spatial distribution of economic activity.

To take an informed stance in this debate, it is useful to recall the notion of spatial transaction costs, or the costs of undertaking and coordinating activities across distance. These are composed of two broad categories: transportation costs and information-transmission costs (McCann and Shefer, 2004). While transportation costs have unequivocally decreased over time, the evolution of information-transmission costs is less straightforward (Iammarino and McCann, 2013).

It is certainly true that the progress of information and communication technologies has enriched organizations' spatial coordination capacity, allowing to orchestrate even highly complex activities from a distance. However, the use of telecommunications does not always substitute for face-to-face interactions. In many situations, the two could be complementary (Gaspar and Glaeser, 1998). As the bulk of knowledge to be transmitted across space is not merely of a codified nature but includes some tacit components, the opportunity costs of coordinating its international movement without leveraging face-to-face contacts increases the volume, variety and complexity of the resulting knowledge (McCann and Shefer, 2004). This suggests that, far from shrinking, spatial transaction costs could even increase with the advancement of communication and transportation technologies, thereby reinforcing the role of geography.

The importance of geography for innovation also keeps on being substantiated by facts. For instance, despite undeniable globalization processes, national and sub-national technological heterogeneity is very resilient (Frost, 2001; Morgan, 2004; Patel and Pavitt, 1994). As suggested by the literature on territorial innovation systems (Lundvall, 1992), countries and regions develop their own patterns of technological specialization, and these seem to be relatively stable over time (Cantwell, 1989). The path-dependency of such evolutionary dynamics depends not only on the ex-ante geographical distribution of technological advantages, but also on the subsequent "localization" of knowledge diffusion (Jaffe et al., 1993).

Though relevant and pervasive, such theoretical and empirical insights should not be interpreted as arguments in support of a direct, unequivocal relationship between proximity and innovation. Clearly, the role of geographical proximity in learning and knowledge creation has to be treated with cautiousness. Recent EG perspectives have suggested that proximity is a multidimensional construct, which can be comprehensively disentangled only by accounting for a combination of organizational, social, institutional and cognitive aspects, in addition to geographical ones (Boschma, 2005).

This approach is fundamental to our analysis, as it clarifies that the rationale behind the role of any of the dimensions of proximity lies in its ability to reduce the uncertainty underlying the innovation process, and to address the coordination problems arising from the need to ensure the exchange of complementary knowledge inputs across heterogeneous agents (Boschma, 2005). While this perspective elucidates that geographical proximity should be considered neither as an essential requirement, nor as a sufficient condition to activate processes of interactive learning and innovation (Boschma, 2005), it simultaneously insinuates that other dimensions of proximity, if considered in isolation, would unlikely allow to overcome the barriers of geographical distance either (Gertler, 2001). In other words, recent approaches that praise for the omnipotent role of organizational or relational proximity could be over-optimistic. Moreover, this perspective endorses the role of geographical proximity as enabler of cooperation and repeated face-to-face contacts, and confirms that – everything else being equal – knowledge will be shared more efficiently among co-located agents.

Face-to-face interaction, in particular, is critical when it comes to overcoming incentive problems, expediting socialization and stimulating individual motivation (Storper and Venables, 2004). The combination of such mechanisms generates the "buzz" that characterizes environments in which people and organizations are able to share information, knowledge and even complex ideas (Bathelt et al., 2004) not because they happen to come into contact, but because they regularly interact among each other thereby building cultural community, shared interpretative models and social affinity (Maskell and Malmberg, 2007).

This is a very straightforward logic, as it suggests that being able to meet and interact in convenient manners, as co-located agents do, is conducive to problem solving and new ideas (Estall and Buchanan, 1961). In other words, local "buzz" is not an outcome of firms' planned behavior, but rather an unintended consequence of the continuous interaction among agents; it does not arise from occasional geographical proximity, but from structural co-location (Maskell and Malmberg, 2007).

On the whole, these reflections suggest that geographical proximity can be intended as the source of a wide-range of opportunities for adaptation, knowledge exchange and innovation; yet, for such opportunities to materialize, other dimensions of proximity may be relevant (Iammarino and McCann, 2013). While this book does not have the ambition to take a clear stand in the recent debate on the role of spatial proximity for knowledge flows (Boschma, 2005; Gertler, 2003; Iammarino and McCann, 2013; Malmberg and Maskell, 2002), it contends that, whether direct or indirect, geographical proximity still plays a vital role in the dynamics of multinational innovation, by influencing internal and external mechanisms of knowledge governance in fundamental ways. Clearly, this does not mean that orchestration of tacit knowledge across distance is impossible, as some MNCs have already demonstrated to have the advanced capabilities to do so (Cantwell and Santangelo, 1999).

It is worth noting here that, if innovation activities are strongly localized, scientific progress is even more geographically concentrated (Florida, 2005). This implies that only a few places in the world are actually able to generate world-class knowledge. Such heterogeneity in the geographical distribution of resources and technological competences makes it clear that there is no single location that – at any given point in time – is able to ensure the availability of the whole set of resources that are necessary to develop the most effective innovations. Organizations that are willing to compete effectively in the global marketplace have to strive to acquire distant knowledge, as this will offer opportunities for value creation that cannot be easily substituted.

This imperative becomes more and more stringent as the pace of technological change increases, making it critical for companies to gain a prompt access to newly created inventions. While recently created knowledge tends to be inherently tacit, going ahead with its life cycle, knowledge may become more explicit thus being easily transferable even without the need of extensive personal interaction. In other words, the exchange of knowledge is likely to be more bounded to geography in the initial stage of knowledge development, which also coincides with the phase in which it has a higher competitive potential, as first-mover advantages can be established. Hence, to be able to benefit from a given invention, it may be critical to be proximate to the invention source in the initial stages of its creation, a condition that is likely to be satisfied only through co-location.

Co-location opportunities represent the relative advantage MNCs possess compared to single-location firms, as they have the chance to overcome the localization of technological activities by establishing their subsidiaries abroad to tap into the heterogeneous pockets of expertise available worldwide. Through the network of foreign subsidiaries, highly valuable knowledge linkages can be built with the diverse and globally distributed local knowledge networks in which MNCs might have an interest (Almeida and Kogut, 1999; Almeida and Phene, 2004), allowing for effective technology sourcing.

Yet, external sourcing of knowledge cannot be considered in isolation from the internal exploitation of such knowledge (March, 1991). It is imperative to acknowledge that while proximity acts as an enabling factor for the acquisition of tacit knowledge from external sources, it also plays an important role in ensuring a smooth management of innovative activities within the firm boundaries. On one hand, because knowledge is sticky and tends to develop particular specialization patterns in specific regions (Jaffe et al., 1993; Markusen, 1996), firms internationalize their R&D to build and leverage a rich external network that may enable them to effectively tap into geographically distant bodies of knowledge and expertise. On the other hand, as MNCs internationalize their R&D function, activities that strongly influence the management of innovation within the firm organizational boundaries, such as control, communication and knowledge sharing become more and more complex, as they have to be orchestrated over long distances.

MNCs can effectively sustain the international organization of R&D activities only if they strive for a delicate trade-off between "external and internal" proximity (Blanc and Sierra, 1999). Not surprisingly, the geographical proximity with the parent company and with the firm's most important manufacturing facilities has long represented the key reason why MNCs have maintained most of their R&D activities at home. Of course, internal and external proximity should not be conceived only in geographic terms, as several other aspects affect the patterns of knowledge sharing and diffusion. Accordingly, MNCs leveraging a network of globally dispersed subsidiaries have developed increasingly effective techniques to counterbalance geographical distance with organizational proximity. Yet, we maintain that geography has fundamentally shaped and will likely continue to deeply affect MNCs' strategic choices in R&D and, in turn, the overall dynamics of global innovative activities.

DOI: 10.1057/9781137555441.0007

1.4 Is R&D really international?

Before exploring the determinants and implications of R&D internationalization (Chapter 2), an aspect that deserves attention is the extent to which MNCs' innovative activities are actually globalized. The growing number of research papers that, starting from the 1980s, have focused on the internationalization of corporate R&D points to the practical relevance of this issue. Yet, some prominent scholars have highlighted that we should be cautious in emphasizing the magnitude of the phenomenon.

Patel and Pavitt (1991) define the technological activities of world's largest firms as an important example of *non-globalization*. Dunning and Lundan (2009) suggest that the internationalization of MNCs' innovative activities has traditionally been limited and has not been able to keep pace with the much more globalized architecture of productive activities despite the significant increase in internationally performed corporate R&D witnessed over the last decades. Recently, Belderbos et al. (2013) observe a persistent dynamic of R&D home-bias, suggesting that MNCs still allocate a disproportionate share of their overall R&D activities to their home-countries. In other words, many authors have suggested that MNCs tend to retain advanced innovative capabilities at home.

The objective of this paragraph is to review the most significant evidence on this phenomenon, adding recent data that may contribute to understand its underlying long-term dynamics. Confronting opposite viewpoints in the literature and bringing fresh information and interpretations on R&D globalization trends help to develop a comprehensive idea on the phenomenon.

As suggested by Dunning and Lundan (2009), there are mainly three empirical approaches to the analysis of the internationalization of innovation: the first, and most common, is based on the use of patent data, the second utilizes survey-based evidence while the third approach employs publicly available information on MNCs' R&D expenditures, and specifically on its geographical distribution.

Starting from the patent-based approach, the first study that has to be mentioned is the one performed by Cantwell (1995b), who analyzed seven decades of patent activity (from 1920 to 1990) by the largest US and European firms, showing that the share of patenting owed to research located abroad increased from 7.91% in 1982–1939 to 14.52% in 1969–1990. This aggregate, slowly rising trend encompasses several more fine-grained dynamics. For instance, while countries such as Germany and France basically followed the average trend, Swedish firms' technological activities were highly internationalized already in the 1920s, although they reduced consistently their global R&D involvement for a significant number of years after the Second World War. A different pattern has characterized countries such as the United Kingdom (UK) and Switzerland, whose R&D internationalization process was initially moderate, but grew to reach a share of foreign patenting structurally superior to 40% starting from the early 1980s. The pursuit of foreign R&D also shows some sectorial distinctions, with the chemical industry spurring the highest share of patenting abroad for the whole period subsequent to the 1940s.

A second relevant analysis of patent data is the study of 686 of the world's largest manufacturing firms conducted by Patel and Pavitt (1991) for the period 1981–1986. Again, the authors document relevant heterogeneity across countries. For instance, while the Dutch companies in the sample conduct 82.0% of their total technological activities abroad, the proportion of foreign innovation realized by the Japanese counterparts is only 0.6%. Moreover, their data show that only for 3 out of the 11 investigated countries, foreign controlled large firms accounted for more than 20% of the national technological activity.

In a subsequent study covering the patenting activity of 569 of the world's largest firms from 13 countries, Patel (1995) observes that, with the exception of Canadian firms, companies had in general increased their proportion of overseas technological activities. In both these studies, the authors conclude that despite the undeniable growing trends, the overall picture is far from representing a situation of globalization of innovative activities.

Though less comprehensive compared with patent-based analyses, the evidence originating from surveys, often compiled by academic researchers or official institutions, offers a complementary and sometimes deeper view on R&D internationalization.

For instance, leveraging information from a survey involving 55 US large companies for the period 1960–1974, Mansfield et al. (1979) show that on average approximately 10% of total R&D expenses was devoted to foreign technological activities, although approximately 75% of foreign

R&D expenses was used for adaptation or improvement of existing products and processes, rather than to entirely new inventions, a percentage which was declared to be much higher compared with that reported for home-country R&D.

In a similar study, Pearce and Singh (1992) found that over 167 firms, approximately 44% had no overseas R&D activities, while nearly 20% devoted one-fifth of their total R&D expenditures to foreign technological activities.

Drawing on survey data from 47 among Japanese and Swedish MNCs, Granstrand (1999) showed that while Japanese firms' R&D internationalization grew faster compared with Swedish counterparts', the foreign share of their technological activities (5%) was much lower than the Swedish one (23%) at the beginning of the 1990s.

A survey of the largest R&D spenders undertaken by UNCTAD in 2004–2005 reported that the average expenditure for foreign technological activities in 2003 was approximately 28% of the total R&D budget. In addition, 69% of the interviewed firms declared to expect their overseas R&D investment to increase, 29% of the companies predicted a stable trend and only a small percentage of 2% hypothesized a reduction of foreign R&D (UNCTAD, 2005).

The third approach that enables to evaluate the internationalization of corporate technological activities relies on data on the share of industrial R&D performed by foreign affiliates of indigenous firms abroad or by domestic affiliates of foreign firms in host countries. To contribute fresh information on the phenomenon, Table 1.1 employs data from the OECD Activities of Foreign Affiliates Database to show the R&D expenditure by foreign affiliates as a percentage of domestic R&D for a set of countries for which data are available in the period 1995–2008. Overall, the trend is an increasing one, as all countries demonstrate a higher percentage in the last available year compared with the first available year. As the above literature reviews suggested, there is strong cross-country heterogeneity, which emerges clearly if we compare for instance the value of Japan (which varies along the whole period between 2% and 3%) to that of Switzerland (which varies along the whole period between 99% and 130%).

Table 1.2 shows the R&D expenditures of foreign firms as a percentage of the total business R&D expenditure of a set of countries for selected years in the period 1995–2011, as well as its variation over selected two-year periods, using data from the OECD Main Science and Technology

TABLE 1.1 Business sector R&D expenditure by affiliates abroad as a percentage of domestic R&D

Year	1995	1996	1997	1998	1999	2000	2001	2002	2003	2004	2005	2006	2007	2008
Country
Germany	17.3	n.a.	17.3	n.a.	18.6	n.a.	27.6	n.a.	23.7	n.a.	29.9	n.a.	24.5	n.a.
Italy	n.a.	n.a.	n.a.	1.7	4.6	3.3	2.3	3	2.8	n.a.	n.a.	n.a.	n.a.	n.a.
Japan	n.a.	n.a.	2.1	n.a.	2.8	2.7	2.5	3.1	2.9	2.9	2.7	2.9	2.8	n.a.
Switzerland	n.a.	114.2	n.a.	n.a.	n.a.	124.1	n.a.	n.a.	n.a.	99.4	n.a.	n.a.	n.a.	131.6
United States	9.5	9.7	9.3	8.7	9.9	10.1	9.8	10.9	11.4	12.4	12.2	11.9	12.8	14.7

Source: OECD Activities of Foreign Affiliates Database.

TABLE 1.2 R&D expenditures of foreign firms as a percentage of total business R&D expenditure, and growth rate

Year	1995	1999	% 1997–1999	2001	% 1999–2001	2007	% 2005–2007	2009	% 2007–2009	2011	% 2009–2011
Country											
Belgium	n.a.	n.a.	n.a.	n.a.	n.a.	59.37	4	53.78	−9	66.00	23
Canada	29.75	32.01	−8	29.64	−7	33.76	4	33.16	−2	34.16	3
Finland	n.a.	14.92	13	14.25	−4	n.a.	n.a.	14.51	n.a.	14.82	2
France	17.09	n.a.	n.a.	21.54	n.a.	20.85	−11	28.08	35	27.45	−2
Germany	16.06	17.84	4	24.77	39	26.21	−6	27.28	4	26.13	−4
Ireland	66.22	63.77	−2	65.23	2	72.37	3	69.86	−3	71.13	2
Japan	1.37	3.93	204	3.39	−14	5.14	0	6.33	23	n.a.	n.a.
Netherlands	n.a.	21.52	5	19.62	−9	n.a.	n.a.	30.18	n.a.	32.51	8
Spain	26.80	32.83	−8	30.99	−6	34.33	31	26.59	−23	35.19	32
Sweden	20.71	36.38	95	40.65	12	34.14	−18	30.93	−9	37.78	22
United Kingdom	29.15	31.16	−5	42.81	37	39.32	0	47.01	20	50.60	8
United States	13.28	13.05	19	13.10	0	15.21	11	14.32	−6	15.36	7

Source: OECD Main Science and Technology Indicators Database.

Indicators Database. Also in this case, the comparison between the value of the first and the last available years reveals an overall increasing dynamic.

Cross-country differences are again very visible. The share registered by Belgium, Ireland and the UK in 2011 is higher than 50% and this is not surprising, as these countries are known for being host to a high number of foreign subsidiaries. The lowest values are associated with Finland and the US, although it is Japan – for which the 2011 value is not available – to account for the lowest share of foreign affiliates' R&D (6.33% in 2009). The remaining countries are placed somewhere in between.

The dynamics that can be detected in the period of analysis are very heterogeneous, although it is evident that countries characterized by the lowest share of R&D expenditures by foreign firms are also those that have experienced the smoothest evolution over time. Moreover, it is useful to detect a possible "global crisis" effect after 2008 (Dachs, 2014). In fact, the variation registered between 2007 and 2009 is negative for 6 out of 10 countries for which data are available.

To complement this picture with recent patent data, we have also reported in Tables 1.3 and 1.4 respectively the share of patents invented abroad and the share of patents owned by foreign residents for selected years, and the relative growth rates, using the OECD dataset on International co-operation in patents, to exploit the seemingly strong coherence between the information provided by R&D evidence and the trends emerging from patent data (Freeman and Hagerdoon, 1995).

For both indicators, the long-term trend is positive, except for some rare cases. For instance, in the case of Singapore, the share of patents invented abroad was 60% in 1980 and 57.21% in 2012. Of course, we should be careful in interpreting these data as the share value measured at the beginning of the period could be driven by a very low absolute number of patents by the registering country. For most of the countries analyzed, both indicators increased by a rate higher than 200%, which even reached the 1300% increase in some cases. For instance, for the US, the share of patents owned by foreign residents increased by a rate of 1380%, although foreign-owned patents represent only the 13% of the whole US patent production in 2012. Not only for countries such as the UK and Switzerland, but also for the most important emerging countries, namely China and India, the indicator is well more than 60% in 2012.

A more significant indication of the recent trend followed by this indicator is probably offered by the variation in the periods 2000–2012 and

TABLE 1.3 Share of patents invented abroad, and growth rate (selected countries and years)

Year	1980	1990	% 1980–1990	2000	% 1995–2000	2009	% 2008–2009	2010	% 2009–2010
Country									
Denmark	8.27	6.36	−23%	21.69	−7%	34.07	−8%	43.84	29%
France	3.62	7.04	94%	21.89	104%	36.74	7%	38.21	4%
Germany	3.71	5.61	51%	13.14	28%	28.49	3%	30.36	7%
Italy	4.09	4.65	14%	9.24	27%	21.15	−6%	21.48	2%
Japan	1.01	1.72	71%	4.28	34%	6.15	5%	6.46	5%
Korea	0.00	2.76	n.a.	4.27	33%	6.00	−16%	6.34	6%
Netherlands	13.54	36.83	172%	57.70	20%	60.50	7%	60.06	−1%
Sweden	4.26	12.58	195%	26.48	20%	43.24	5%	47.34	9%
Switzerland	23.51	29.03	23%	51.92	29%	66.51	2%	67.58	2%
United Kingdom	6.27	11.10	77%	19.51	3%	40.48	12%	42.87	6%
United States	5.58	6.27	12%	9.57	18%	14.81	−7%	14.72	−1%
Brazil	10.00	2.50	−75%	7.89	−47%	41.80	14%	37.41	−10%
China (People's Republic of)	33.33	12.50	−62%	26.13	5%	32.12	−12%	32.07	0%
India	0.00	6.25	n.a.	3.50	19%	45.60	−1%	48.14	6%
Singapore	60.00	0.00	−100%	34.40	48%	61.23	5%	61.46	0%
South Africa	7.50	6.25	−17%	18.45	44%	29.21	−9%	22.39	−23%
European Union (28 countries)	2.31	4.21	82%	12.29	42%	25.27	6%	28.45	13%
OECD – Total	4.95	5.75	16%	10.56	31%	13.72	−4%	13.85	1%
World	5.11	5.81	14%	10.83	33%	14.82	−3%	15.17	2%

Source: OCED dataset on International co-operation in patents.

TABLE 1.3 Continued

Year	2011		2012				
Country		% 2010–2011		% 2011–2012	1980–2012	1990–2012	2000–2012	2005–2012
Denmark	51.27	17%	52.07	2%	530%	718%	140%	75%
France	44.62	17%	52.90	19%	1360%	652%	142%	81%
Germany	32.67	8%	35.00	7%	842%	524%	166%	68%
Italy	25.99	21%	35.41	36%	765%	661%	283%	130%
Japan	6.33	−2%	6.73	6%	569%	292%	57%	37%
Korea	5.97	−6%	8.86	49%	n.a.	222%	108%	54%
Netherlands	60.51	1%	57.25	−5%	323%	55%	−1%	0%
Sweden	47.16	0%	50.00	6%	1073%	297%	89%	43%
Switzerland	70.60	4%	74.59	6%	217%	157%	44%	27%
United Kingdom	44.21	3%	44.59	1%	611%	302%	129%	48%
United States	14.48	−2%	13.97	−4%	150%	123%	46%	6%
Brazil	44.87	20%	48.48	8%	385%	1839%	514%	62%
China (People's Republic of)	36.26	13%	37.53	3%	13%	200%	44%	8%
India	50.81	6%	53.08	4%	n.a.	749%	1417%	45%
Singapore	64.91	6%	57.21	−12%	−5%	n.a.	66%	−9%
South Africa	39.47	76%	45.45	15%	506%	627%	146%	90%
European Union (28 countries)	32.48	14%	34.16	5%	1380%	711%	178%	97%
OECD – Total	13.94	1%	14.22	2%	187%	148%	35%	11%
World	15.45	2%	15.72	2%	207%	171%	45%	18%

Source: OCED dataset on International co-operation in patents.

TABLE 1.4 Share of patents owned by foreign residents, and growth rate (selected countries and years)

Year	1980	1990	1980–1990	2000	1995–2000	2009	2008–2009	2010	2009–2010
Country									
Denmark	19.23	16.27	−0.15	31.02	−0.03	45.62	−0.12	49.01	0.07
France	9.15	13.98	0.53	28.54	0.40	40.98	0.00	45.09	0.10
Germany	10.43	11.43	0.10	19.07	0.15	35.79	0.05	38.18	0.07
Italy	10.88	15.57	0.43	27.46	0.14	41.29	0.00	38.77	−0.06
Japan	1.49	1.75	0.18	3.30	−0.05	4.94	−0.02	4.92	0.00
Korea	0.00	4.41	n.a.	5.43	0.15	5.28	−0.05	6.56	0.24
Netherlands	58.29	57.57	−0.01	40.50	−0.34	51.52	0.01	53.29	0.03
Sweden	7.35	18.40	1.50	18.54	−0.10	38.11	0.12	45.35	0.19
Switzerland	30.83	34.61	0.12	40.66	−0.03	62.50	−0.02	60.54	−0.03
United Kingdom	22.12	29.06	0.31	51.85	0.29	64.13	0.04	65.05	0.01
United States	0.87	1.81	1.08	5.54	0.76	13.93	−0.02	13.95	0.00
Brazil	38.71	15.22	−0.61	53.29	0.10	58.82	0.21	59.42	0.01
China (People's Republic of)	0.00	40.38	n.a.	65.67	0.20	65.13	0.01	63.64	−0.02
India	75.00	60.53	−0.19	54.87	−0.10	81.34	−0.03	86.68	0.07
Singapore	71.43	83.33	0.17	51.30	−0.22	56.85	−0.01	60.49	0.06
South Africa	16.67	30.56	0.83	32.61	0.09	46.51	0.06	39.39	−0.15
European Union (28 countries)	13.23	14.80	0.12	21.44	0.03	32.75	0.03	36.61	0.12
OECD – Total	5.03	5.65	0.12	10.29	0.31	13.89	−0.04	14.02	0.01
World	5.11	5.81	0.14	10.83	0.33	14.82	−0.03	15.17	0.02

Source: OCED dataset on International co-operation in patents.

TABLE 1.4 Continued

Year	2011			2012		
Country	2010–2011	2011–2012	1980–2012	1990–2012	2000–2012	2005–2012
Denmark	57.38	0.17	60.00	0.05	212%	269%	93%	68%
France	51.41	0.14	63.26	0.23	591%	353%	122%	73%
Germany	41.68	0.09	46.12	0.11	342%	303%	142%	85%
Italy	46.78	0.21	58.05	0.24	433%	273%	111%	88%
Japan	4.65	–0.06	5.92	0.27	297%	237%	79%	54%
Korea	6.25	–0.05	7.74	0.24	n.a.	76%	43%	90%
Netherlands	59.40	0.11	51.31	–0.14	–12%	–11%	27%	34%
Sweden	50.20	0.11	53.31	0.06	625%	190%	188%	92%
Switzerland	67.82	0.12	73.54	0.08	139%	112%	81%	53%
United Kingdom	71.14	0.09	72.21	0.02	226%	149%	39%	30%
United States	13.92	0.00	12.88	–0.07	1380%	612%	133%	32%
Brazil	75.64	0.27	71.05	–0.06	84%	367%	33%	41%
China (People's Republic of)	64.30	0.01	60.47	–0.06	n.a.	50%	–8%	–7%
India	87.31	0.01	87.02	0.00	16%	44%	59%	22%
Singapore	62.42	0.03	66.67	0.07	–7%	–20%	30%	16%
South Africa	55.26	0.40	72.73	0.32	336%	138%	123%	98%
European Union (28 countries)	43.34	0.18	46.43	0.07	251%	214%	117%	95%
OECD – Total	14.41	0.03	14.92	0.04	196%	164%	45%	19%
World	15.45	0.02	15.72	0.02	207%	171%	45%	18%

Source: OCED dataset on International co-operation in patents.

DOI: 10.1057/9781137555441.0007

2005–2012. It can be observed that the US has experienced an increase respectively of 133% and 32% in the two periods, while the European Union (28 countries) has registered growth rates of 117% and 95%. As far as the share of patents invented abroad is concerned, the trend is less pronounced for the US, which over the two periods registers a variation respectively of 46% and 6%, and stronger for the European Union (178% and 97%). For several countries, the value of the indicator in 2012 is equal or higher than 50%, reaching almost 75% for Switzerland. As noticed from the data on R&D expenditures of foreign firms as a percentage of the total business R&D expenditure, also patent data hint at potential global crisis effects, with negative variations for several countries in both the share of patents invented abroad and the share of patents owned by foreign residents between 2008 and 2009.

On the whole, this analysis suggests that the extent of R&D internationalization varies across countries, industries and companies (UNCTAD, 2005; von Zedtwitz and Gassmann, 2002). As highlighted by Niosi and Godin (1999), firms from small countries such as Sweden, Belgium and Switzerland tend to rely heavily on foreign R&D. Conversely, in large industrialized countries, there are both intermediate solutions of moderate R&D internationalization, as in the case of French, German and American companies, and more extreme solutions, as in the case of Japanese companies that are persistently more focused on domestic technological activities. The phenomenon is also critically influenced by conditions in the home- and the host countries, as well as by MNCs' resources and strategies (Belderbos et al., 2008, 2013; Narula, 2002; Pearce, 1989), as Chapter 2 explains.

Though it is undeniable that R&D and innovation have long lagged behind other firm activities in terms of internationalization intensity, we contend that the dynamics of this phenomenon, which have been depicted from several viewpoints and through several data sources, portray a very clear process of geographic dispersal, aided by the increasing improvements in communication technology as well as by the accumulation of MNC experience in developing and perfecting group-level routines that ease the coordination of decentralized R&D to create global networks. This process is even more manifest if considered in the light of the inherent complexities associated with the management and the organization of value-creating activities, such as those involving knowledge and innovation, compared with more standardized activities. As many authors have suggested and as next chapter shows, a deeper investigation into the *type* of innovative activities that MNCs internationalize is needed to understand whether the records on R&D internationalization should point to a qualitative evolution, rather than to a mere quantitative dynamic (Belderbos et al., 2013; Florida, 1997). In other words, to evaluate the relevance of R&D globalization, it is important to ascertain whether the R&D activities performed abroad are predominantly adaptive, or even more creative tasks are executed outside the MNC home-base.

Notes

1   The R&D spending of the top 20 R&D spenders in 2011 ranged between the 9.9 billion dollars spent by Toyota and the 5.5 billion dollars spent by AstraZeneca (Booz & Company, 2012), while the GERD of the mentioned countries ranged between the 5.1 billion dollars spent by Norway and the 2.7 billion dollars spent by Hungary (data extracted in 2015 from OECD. Stat).
2   These MNCs are GlaxoSmith-Kline, Honda, IBM, Intel, Johnson & Johnson, Microsoft, Novartis, Pfizer, Roche, Samsung, Sanofi, Toyota and Volkswagen (Jaruzelski et al., 2014).

2
Managing Innovation in Multinational Firms

Abstract: *As the breadth and sophistication of foreign R&D activities increase, the complexity of organizing geographically distributed R&D units augments. Headquarters have the power and the responsibility to initiate the process of R&D internationalization, to guide its evolution and to orchestrate the MNC's geographically distributed R&D activities.*
This chapter reviews and discusses the literature on MNC headquarters' decisions regarding the geographical dispersal of MNC innovative activities. First, it analyses centripetal and centrifugal forces that influence the choice to internationalize R&D. Second, it describes the different typologies of FDI in R&D focusing on their mandated objectives and activities. Third, it systematizes the factors influencing location choices. Finally, it focuses on the organizational arrangements that can be deployed to effectively orchestrate the R&D geographical dispersal.

Keywords: autonomy; centripetal and centrifugal forces; coordination; location choices; MNC headquarters

Perri, Alessandra. *Innovation and the Multinational Firm: Perspectives on Foreign Subsidiaries and Host Locations.* Basingstoke: Palgrave Macmillan, 2015.
DOI: 10.1057/9781137555441.0008.

2.1 Forces behind R&D internationalization

Although the purpose of this book is to enrich our understanding of research and development (R&D) internationalization by leveraging perspectives on subsidiaries and host locations, ignoring the bulk of research that along the years has scrutinized headquarters-level decisions regarding the geographical dispersal of multinational corporations (MNCs) innovative activities would be inappropriate. In fact, it is the headquarters to choose whether to establish an R&D subsidiary abroad, to possibly identify a location and to offer strategic guidelines on how to conduct local innovation activities.

In other words, despite the critical role that subsidiaries and host locations may play in determining the performance and evolution of foreign R&D, headquarters have the power to initiate the process of R&D internationalization and to affect its development over time. Hence, their strategies and the factors influencing their decisions need to be fully acknowledged, as they are primary aspects with which subsidiaries and geographies interact.

The first major choice headquarters need to take, and that will constitute the focus of this paragraph, is whether to internationalize the MNC innovation activities. In this regard, decades of literature have concluded that there are two sets of forces, which can be distinguished based on the contrasting effects they exert on the decision to internationalize innovation. In fact, while there are factors that drive MNCs to geographically centralize R&D activities in the firm home-country, other forces encourage the evolution toward the geographical decentralization of innovation (Granstrand et al., 1993; Hirschey and Caves, 1981; Pearce, 1989).

The first set of forces, which have been labeled *centripetal forces*, can be categorized in the following four factors:

1 scale and scope economies in R&D;
2 coordination costs;
3 protection of firm-specific technology;
4 embeddedness in the domestic innovation system.

As far as the *centrifugal forces* are concerned, these can be broadly distinguished between:

1 demand-driven forces;
2 supply-driven forces.

Starting from the first set of forces, it is useful to recall the work by Hewitt (1980) and Hirschey and Caves (1981) who suggest that a first driver of R&D centralization is the prospect to realize scale and scope economies. While significant differences may exist across industries and technologies (Kuemmerle, 1998), R&D tends to be partly indivisible in its nature. To perform R&D activities, firms need to invest into expensive and specialized equipment (i.e., machinery and laboratories) and scientific expertise, which require a minimum threshold volume to be economically feasible. Spreading R&D labs in different locations entails a replication of idiosyncratic assets and human capital investment, leaving resources largely under-utilized.

Along with scale economies, centralized corporate R&D activities may also generate scope economies arising from the knowledge spillovers that may potentially occur among different technological fields (Arora et al., 2011; Henderson and Cockburn, 1996). On one hand, knowledge is increasingly interdisciplinary and "universal", meaning that it may be deployed in contexts that are different from those for which it was originally developed (Arora and Gambardella, 1994). On the other hand, MNCs have gradually undergone a process of technological diversification (Zander, 1997). Accordingly, centralized R&D activities offer a favorable context to realize synergies from sharing and reusing knowledge across diverse technological fields.

A second force leading to centralization lies in the high coordination costs of distributed R&D activities. R&D investments are fully exploited only when their outcomes can be employed by the entire organization. Yet, knowledge and innovative assets encompass some tacit components whose transfer, even within the boundaries of the same firm, requires extensive interaction, timely communication and action, and effective learning among the firm R&D facilities. These mechanisms are facilitated when scientists of different R&D units (and their counterparts in other relevant corporate functions, such as top-management, production and marketing) are located in the home-country, where coordination needs can benefit from the advantages of geographic proximity. Conversely, geographical distance – especially when coupled with cultural distance (Sosa et al., 2002) – comes along misunderstandings, lack of trust and reduced opportunities for informal technical discussions, thus hindering fluid information exchange and leading to a substantial increase in coordination costs (von Zedtwitz and Gassmann, 2002).

Another reason for keeping R&D at home is related to the protection of the firm-specific technological assets. MNCs are strongly committed with the appropriation of the outcomes of their innovative efforts. Yet, when conducting R&D abroad, proximity to foreign competitors activates a wide range of potential channels for spillover (Alcácer and Chung, 2007; Alcácer and Zhao, 2012; Shaver and Flyer, 2000). The geographical distribution of R&D laboratories multiplies the sources of knowledge leakage and makes effective protection of the firm's technological resources on the part of corporate headquarters more difficult to achieve. Conversely, when innovative activities are centralized in the home country, MNCs may more easily ensure their control over knowledge-based assets.

It should also be noted that this centripetal force is related to the previously analyzed force. In fact, the higher degree of internal coordination and communication required by a global distribution of innovative activities often comes along a process of knowledge codification that enables technological knowledge that was originally tacit to be transferred across distance despite the absence of personal interaction. Clearly, codified knowledge spanning national boundaries to be shared with foreign MNC units is much more exposed to the risk of external dissemination compared to tacit knowledge that is concentrated at home and carefully protected by the headquarters' central control (Rugman, 1981).

The facilitated protection of the firm's technology in presence of R&D domestic concentration may also be related to headquarters' tendency to centralize intellectual property management in the home-country. When R&D activities are performed in close coordination with intellectual property (IP) specialists, the full appropriation of intangible assets is more feasible. Domestic IP management routines tend to be reliable and sophisticated, thereby ensuring a wider protection of innovative outcomes. Conversely, the lack of appropriate IP practices at the MNC's peripheral locations may hinder effective protection and commercial exploitation of the firm knowledge. This idea has been conceptualized through the notion of "safe nests" (Di Minin and Bianchi, 2011), used to identify those domestic R&D centers that ensure high levels of appropriability of firms' R&D outcomes. MNCs that concentrate their IP management system at home, thus generating an imbalanced distribution of appropriability levels across the MNC network, are more likely to carry out their most promising innovation projects in the domestic *safe* nests as these ensure a more thorough exploitation of potentially strategic technological resources (Di Minin and Bianchi, 2011).

A final critical centripetal force originates from the relationship between the MNC and its home-country. Firms that have traditionally organized their R&D activities in the home-base face a "structural inertia" phenomenon (Narula, 2002), which leads them to keep core R&D projects at home in order to stick to established R&D hubs. Sunk investments and the legacy of past choices are among the potential factors that explain such path dependent logic in the allocation of R&D resources. MNC advantage is often rooted in the extensive web of connections with actors in the local business environment, and particularly with customers and suppliers. These connections often instigate a co-evolution process through which technological capabilities become increasingly interdependent (Freeman, 1987; Lundvall, 1992; Nelson, 1993), thereby shaping the nature of the innovation process within the firm. The embeddedness in the home-base innovation system may therefore act as a critical factor that MNCs leverage to sustain their technological strengths, since the persistent interaction with the local business network grants them an insider status along with an easy access to relevant knowledge. Conversely, establishing R&D subsidiaries in foreign locations does not immediately allow to source valuable information, as it takes time and resources to build-up cooperative and trust-based relationships through which tacit knowledge flows. In other words, while the home-country network is easy to maintain, developing new relationships for knowledge sourcing abroad is much more costly and less automatic (Belderbos et al., 2013), although it should be noted that repeated knowledge sourcing in the same system of innovation may generate lock-in situations that undermine the value of innovative outcomes (Narula, 2002).

Along with centripetal forces, there is a set of competing centrifugal drivers that pull R&D outside a MNC's home-country. These can be broadly distinguished between those that emanate from the demand-side and those that proceed from the supply-side.

Demand-driven forces mainly address a need for market proximity and reflect the call for adaptation arising from the establishment of foreign production facilities that aim at enriching the value of centrally developed technology for local customers. Foreign production unavoidably requires some technological adjustments, as both products and production processes need to be tailored to local practices and demand expectations. Local R&D units may be created to comply with these requirements, and often become responsible of further technical responsibilities, such as service and customization (Granstrand et al., 1993). From the headquarters perspective, this type of foreign R&D facilities has a narrow effect on the firm's R&D organization: because the activities performed abroad are specific to a given geographical market and typically limited in their scale, efficiency, coordination and controls are minor issues.

As highlighted in Chapter 1, scholars have gradually realized that demand factors are not the only critical drivers of R&D internationalization. With the emergence of new centers for world-class research and innovation, the supply-side became increasingly important thereby acting as a powerful centrifugal force. While until the end of the 1970s the United States (US) and some areas of Western Europe exerted an almost exclusive dominance in most technological fields, new locations in the "triad" countries became gradually very active in the development of leading-edge knowledge in a broad spectrum of technical areas (Gerybadze and Reger, 1999).

Following these changes, innovative actors – and particularly the leading innovative enterprises – recognized that maintaining a technological leadership requires access to a variety of scientific and technical inputs that cannot be found only in their home-country or in single host locations (Kuemmerle, 1997). It should be noted that this principle is true not only for firms that are willing to expand in new and previously unexplored technical fields. As new technologies are increasingly complex and interdependent, companies need to widen their knowledge base to support the development of their technological competence even in areas where they already possess a technological advantage (Cantwell and Piscitello, 2000).

Overall, the changing competitive dynamics, which in the last decades have evolved particularly in terms of increasing speed and scope of technical change, have ascribed a relevant value to the diversity that can be found in multinational contexts (Zanfei, 2000). Accordingly, along the last decades, an increasing number of countries and sub-national areas have emerged as generators of relevant knowledge, highly skilled R&D personnel and innovative activities.

In this renewed situation, MNCs have realized the importance of delegating to foreign subsidiaries the task of scrutinizing distant locations as possible sources of front-line technologies and of accessing distant knowledge to strengthen their home-base technological competences.

Subsidiaries therefore act as distance-bridging engines that exploit their geographical proximity with foreign sources of technology to perform activities such as knowledge exploration and knowledge access to nourish the MNC existing knowledge-base. Contrary to the technological dependence of adaptive R&D units, this new type of R&D subsidiaries can engage in pre-competitive basic or applied research as well as in original product development.

2.2 Types of innovation-driven foreign direct investment

The analysis of the centrifugal forces behind R&D internationalization suggests that the types of R&D activities MNCs perform abroad are not homogeneous. In particular, there is a profound qualitative difference between the supply-driven R&D unit and the traditional adaptive R&D subsidiary model (Cantwell, 1995b; Florida, 1997). The essence of this difference lies in the creative role assigned to the former, which is demanded to develop new areas of capabilities that could augment the core technological competences of the parent company by leveraging knowledge assets embedded in the foreign location.

It is precisely the diffusion of such creative roles that makes MNC innovation authentically international (Cantwell, 1995b). In fact, although scholars agree in considering both the global *exploitation* of technology and the global *generation* of technology as critical components of the *technological globalization* phenomenon (Archibugi and Michie, 1995; Kuemmerle, 1999), it is possible to claim that while foreign R&D activities that are adaptive and market-oriented fall within those "global localization strategies" in support of foreign manufacturing and product development (Porter, 1986, 1990), MNCs' innovation becomes unequivocally global only with the emergence and gradual diffusion of *technology-oriented, creative overseas R&D activities*. In such cases, headquarters' choice to move R&D activities abroad is dependent not on the internationalization patterns of other MNC functions, but rather on R&D-specific, clear strategic intents.

Along the years, scholars have tried to identify the variety of R&D unit configurations used by MNC headquarters to organize their geographically dispersed innovative activities based on several criteria such as the underlying motives, nature, geographical scope and level of technological competence. To offer an overview of such wide-ranging R&D subsidiary roles, the following paragraph reviews the most relevant classifications that have been proposed by both conceptual and empirical research, revealing a substantial degree of coherence among the proposed types (Figure 2.1).

One of the earliest comprehensive taxonomy of foreign R&D units has been proposed by Håkanson and Nobel (1993) along the lines of the work of Pearce (1989). This empirically generated classification identifies the following five categories of overseas R&D units:

- *Market-oriented units.* These units are fully motivated by market proximity, as they are needed to adapt centrally developed technology to the specificities of the local environment, such that it can be fully exploited abroad as it happens in the home-country. Typically, they are involved in activities such as technological support, customization and service;
- *Production support units.* This type of R&D unit is typically found in association with manufacturing activities that are unique in the MNC network, as they are only developed in a particular location. Accordingly, locally performed R&D activities are highly specialized in offering support to the specific production line. Such R&D types may arise from the pursuit of a product diversification strategy, but they can also be used as a consequence of global product mandates or international production rationalization plans;
- *Research units.* Contrary to the previous types, this category of R&D unit is located in a foreign country irrespective of the simultaneous presence of manufacturing facilities. This sort of "independence" from production activities has been identified as the signal of an entirely different strategic intent, compared to previously analyzed R&D units. In this case, the objective of the overseas investment is the acquisition of specialized technological knowledge embedded in the host country, either through the interaction with other firms, universities or research centers, or through the employment of skilled local human capital;
- *Politically motivated units.* This is a special case of "forced" R&D foreign activities. When foreign direct investment (FDI) occurs by the means of acquisitions to access to strategic assets such as brands or distribution capacity, MNCs are required to be responsive to

host governments' requests. Therefore, they often negotiate to maintain the associated local R&D unit, even if it represents a duplication of the MNC's existing activities. Conversely, when the target's R&D laboratory represents the very motivation for the foreign acquisition, a condition that is increasingly frequent for emerging markets' investors, MNCs are highly willing to preserve and cultivate foreign R&D facilities to magnify the acquisition's benefits (Granstrand et al., 1993) through the access to existing and well developed technological capabilities;

- *Multi-motive units.* These laboratories are very peculiar because they tend to carry out high-level R&D activities for a variety of reasons and in light of several responsibilities, as it happens in central innovative units.

While the abovementioned classification is more comprehensive, Graham (1992) focuses only on technology-oriented R&D initiatives and suggests the existence of two different types of laboratories:

- the first type covers the so-called *listening posts,* that is, foreign labs whose objective is to scrutinize the evolution of technology in the foreign location, in order to capture any relevant trend that the MNC may explore further and potentially engage with. This objective is usually achieved through the development of local knowledge linkages that may serve as information-gathering tools (Almeida and Phene, 2004), useful to gain insights on the research trajectories local actors pursue within distant regions;
- the second type, namely the *generating station,* takes a more active role in creating new knowledge starting from the distinctive host country technology and leveraging interacting relationships with the local scientific and technological community.

Adopting a more general approach, Kuemmerle (1997) develops a classification that has gained great diffusion in subsequent IB studies, distinguishing between *home-base exploiting* and *home-base augmenting* R&D FDI. This classification is based on the canonical distinction between foreign labs whose establishment is pulled by country-specific differences in customer needs, work practices, distribution and supply availability, and R&D units that are instead motivated by the willingness to tap into a specialized knowledge-base, that can be leveraged to enhance the existing MNCs' technological core competences, thereby sustaining its long-term development.

Iwasa and Odagiri (2004) propose a similar distinction, separating *research-oriented* from *local-support-oriented* foreign R&D units. They find that the host location technological strength is a critical determinant of innovation performance only for research-oriented subsidiaries, for which local knowledge sourcing is a relevant responsibility.

Consistent with previous models, Nobel and Birkinshaw (1998) distinguish among three major typologies:

- the *local adaptor* R&D subsidiary, which is consistent with the support-oriented units identified by Pearce (1989) and Håkanson and Nobel (1993), as it has a merely local geographic scope and plays a major role in ensuring the adaptation of headquarters' technology to the specificities of the local market;
- the *international adaptor* R&D subsidiary, which has a broader geographic scope along with a more creative role, as it involves in product development for international markets to adjust to the emergence of globalized manufacturing;
- the *international creator* R&D subsidiary, which is consistent with Håkanson and Nobel's (1993) *research unit* type, in that it is located in a specific foreign country but operates independently from any local production activity. It plays an entirely creative role in both upstream and downstream innovative activities, in direct cooperation with corporate and divisional R&D units.

Rather than identifying distinct R&D unit typologies, other scholars have discussed the wide range of activities R&D subsidiaries may perform abroad. In this regard, Florida (1997) offers an interesting set of findings based on a survey of foreign-affiliated R&D laboratories in the US that essentially point to a prevailing *technology-orientation posture*. In fact, he reports that the three most important activities the interviewed affiliates declared to perform consist in (1) generating new product ideas, (2) gathering information on both technological and scientific advances, and (3) sourcing highly skilled human resources to involve in both upstream and downstream stages of the innovation process (Florida, 1997).

Similarly, Dunning and Lundan (2008) develop a conceptual taxonomy of responsibilities of overseas R&D affiliates, which include the following activities:

- activities involving the adaptation of *products, materials or processes*, and covering tasks and responsibilities that pretty much resemble those described for Håkanson and Nobel (1993)'s *market-oriented units*;
- research activities on *basic materials or products*, which are usually relocated abroad in case of better availability of specific materials in host countries, or when the proximity to the market is critical in the very first stages of the product development;
- *efficiency-seeking research* activities, conducted in locations endowed with particularly advantageous innovation infrastructures, with the aim of redeploying potential outcomes in other parts of the MNC organization;
- research activities to source *technological assets or capabilities*, which by and large coincide with the activities performed by the *research unit* type identified by Håkanson and Nobel (1993).

Beyond surveying the diverse roles and responsibilities R&D subsidiaries may be expected to play in host locations, many of the reviewed studies reveal a trend of geographical separation between different types of R&D labs. Given the subsidiary-centric and location-centric focus

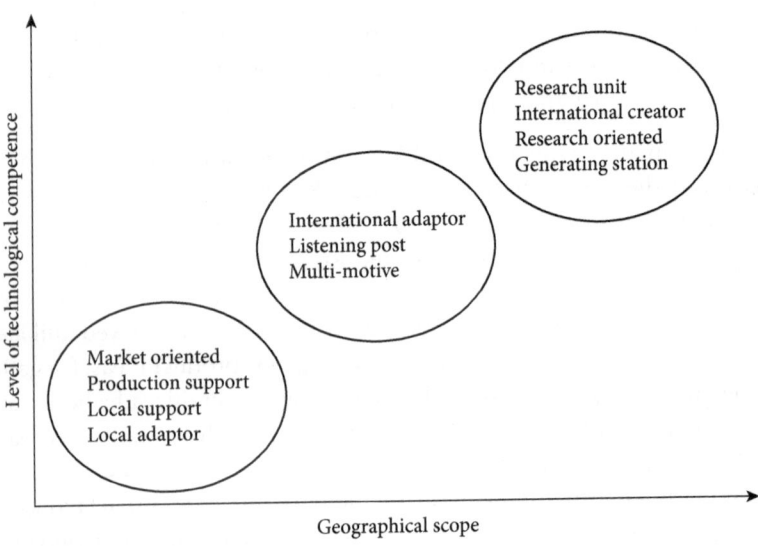

FIGURE 2.1 *Type of foreign innovative units by geographical scope and level of technological competence*

of this study, this is a critical finding. In fact, because in principle it should be much easier for MNCs to concentrate the activities of exploiting and augmenting home-base knowledge in the same R&D affiliate, the decision to locate different types of labs in different locations is a strong evidence of the existence of entirely heterogeneous locational drivers (Kuemmerle, 1999). Location heterogeneity is therefore a strong driving force of MNC choices, even when compared with the potential diseconomies associated with geographical separation between different R&D affiliates.

2.3 Location choices of R&D facilities

Location choices represent the most immediate reflection of firms' geographic behavior. Yet, as both IB and EG scholars have suggested, research into the location strategies of MNCs has not developed with the same degree of comprehensiveness and accurateness that have been devoted to the study of the other two building blocks of the eclectic paradigm, namely ownership and internalization advantages (Beugelsdijk et al., 2010; Beugelsdijk and Mudambi, 2013; Iammarino and McCann, 2013). While this issue is addressed more extensively in Chapter 4, the following text seeks to systematize the most relevant studies that have dealt with MNC location choices of R&D activities.

Clearly, choosing where to locate an R&D facility encompasses a range of idiosyncratic factors that make such decisions particularly complex. Innovation processes are by definition inherently uncertain; hence, adding the multinational dimension further increases the risk involved in these activities. MNC headquarters are therefore put under a great deal of pressure when taking R&D location choices.

It is the firm higher management, often advised by the R&D and strategy departments, to be responsible for the choice of where to establish foreign R&D units. Generally, it is possible to argue that this choice is usually made in consideration of four sets of factors (von Zedtwitz and Gassmann, 2002):

- ▸ the expected output (in terms of market access, knowledge sourcing opportunities, collaboration with the local scientific system, etc.);
- ▸ the quantity and quality of science and technology supply in the location (availability of skilled R&D human resources, sophistication of the local scientific infrastructure, etc.);

- the general operating efficiency of the R&D investment (scale of the R&D unit, cost implications, existence of synergetic effects with the R&D internal network, etc.);
- contingency factors (political and social profile of the location, tax structure, etc.).

A general principle on which scholars have concurred is that location choices will differ depending on whether the foreign R&D activity is driven by home-base-exploiting or home-base-augmenting motivations (Kuemmerle, 1997). Home-base-exploiting R&D tends to follow the MNCs' overseas production and sales thereby being attracted by market-specific characteristics. Home-base-augmenting R&D needs to be placed in areas that are rich in scientific resources and where the local technological effort is high. This is due to the role of spillover mechanisms, which have been identified as a fundamental driver of R&D internationalization.

Knowledge spillovers occur when localized knowledge leaks, either intentionally or unintentionally, beyond innovative actors' organizational boundaries, and can be internalized and used by co-located agents. Spillovers may be of two kinds: *intra-industry spillovers* arise from the co-location of specialized activities pertaining to the same industry (Marshall, 1920), while *inter-industry spillovers* occur when diverse technological activities concentrate in the same geographical area (Jacobs, 1969). Both intra-industry and inter-industry spillovers represent strong attractors for MNCs' foreign investment in R&D (Cantwell and Piscitello, 2005): the former enable to gain access to the latest advances in a firm's area of specialization that ripen in foreign locations, while the latter expose to broad and diversified technology, thus favoring the emergence of new knowledge across sectors. Spillovers arising from a location's scientific infrastructure also influence MNCs' selection of the most adequate location for their R&D activities (Cantwell and Piscitello, 2005). In fact, industrial knowledge creation benefits from the support of a wide-range of external innovative actors, such as research centers, universities and the education system.

Inspired by the distinction between home-base-exploiting and home-base-augmenting motivations, early research on R&D location choices has mainly explored whether the target country selection was more driven by the technological strengths of the home-country or by the knowledge-based assets available in the host country. This was a relevant question to be answered in the 1990s, as scholars were still seeking evidence on the magnitude of technology-driven foreign R&D investment. A prominent role of the host country technological endowment would have indirectly confirmed the increasing diffusion of this type of R&D activities, all else being equal.

Different location strategies can be identified depending on the technological strengths and weaknesses MNCs enjoy at home, vis-à-vis those they could experience abroad (Le Bas and Sierra, 2002):

1 *technology-seeking FDI in R&D* seek to counterbalance MNC domestic disadvantages in specific technological areas by locating in host countries that excel in such fields;
2 *home-base-exploiting FDI in R&D* are triggered by the willingness to leverage a technological strength that the MNC enjoys at home in a foreign country that is relatively disadvantaged in the same technology;
3 *home-base-augmenting FDI in R&D* aim at enriching the advantages a MNC possesses in a given technological area in the home-country by entering host countries that are also relatively strong in the same fields;
4 *market-seeking FDI in R&D* are usually undertaken in technological fields in which the MNC is relatively weak at home, and tend to target host countries that are also technologically deficient in the specific area of investment, often with the objective to access a new market by means of acquisitions.

According with this model, while technology-seeking and home-base-exploiting FDI in R&D presuppose the existence of a technological asymmetry between the home and the host country, home-base augmenting and market-seeking FDI in R&D occur between countries that share approximately the same degree of technological strength.

In addition, home-base-exploiting and home-base-augmenting FDI in R&D identify two very different, although not mutually exclusive, routes to manage a home-base advantage: the first type of FDI in R&D encompasses the leverage of such advantage in a foreign country that is relatively weak in the focal technical area and is based on a *myopic learning* logic that takes place by merely adapting technologies and products to the specificities of the overseas market; the second type of FDI in R&D aims at strengthening the firm's existing technological competence by complementing it with new but related capabilities that are available in technologically rich host countries, thus pursuing a *dynamic learning* logic that is oriented to the long-term preservation and enhancement of the MNC knowledge base (Le Bas and Sierra, 2002).

The empirical assessment on the diffusion of the different location strategies has long offered evidence on the dominant role played by firms' home-country technological advantage in guiding foreign R&D location choices. For instance, patent data analyses suggest that most of the largest MNCs involving in foreign R&D in the period from the late 1980s to the late 1990s invested in technological fields in which they enjoyed an advantage at home (Le Bas and Sierra, 2002; Patel and Vega, 1999). Yet, among these cases, the home-base-augmenting strategy seems to be the most widespread, thus pointing to a predominance of the dynamic learning approach (Le Bas and Sierra, 2002). Moreover, the largest increase in foreign R&D was registered in fields featuring a strong complementarity between the innovative activities performed by the MNC at home and the host country's technological profile (Patel and Vega, 1999).

The importance of the country of origin's technological endowment was also confirmed and further extended in a famous piece by Cantwell and Janne (1999), in which the authors propose a locational hierarchy model according to which MNCs originating from the most advanced locations in their industry are more likely to pursue a complex technological division of international innovative labor, by separating diverse fields of technological development across space. Following this approach, MNCs from higher-order regions tend to focus on distinct technological fields in different host locations. This enables them to complement their existing home-base advantages by leveraging heterogeneous pockets of specialized knowledge that are distributed worldwide. Conversely, MNCs emanating from weaker locations in an industry tend to evolve toward a domestic replication model, thereby reproducing abroad the same trajectory of their home-country technological specialization. The rationale behind this model lies in the uneven endowment with innovation skills and absorptive capacity that companies from leading and laggard locations possess: contrary to firms from lower-order innovation centers, leading country firms are not only more able to innovate by leveraging locally available resources, but are also better positioned to share the newly created knowledge within the MNC network, due to the MNC's higher absorptive capacity.

On the whole, these works point to the critical role played by the home-country's technological profile in shaping MNCs' R&D geographical behavior and offer limited support for the idea that MNCs go abroad to compensate for their domestic technological weaknesses. Yet, some notable exceptions exist, particularly in more recent empirical studies. For instance, leveraging the rich information included in patent data, both Almeida (1996) and Singh (2007) find that MNCs carry out an extensive activity of knowledge sourcing in host countries that are relatively more advanced, most likely to offset a lack of specific technological capabilities at home. The "catch-up" motive is evident also in historical data highlighting that MNCs from small countries such as Switzerland, the Netherlands and Sweden (such as Novartis, Hoffmann-La Roche, Phillips and Ericsson) have forged the process of R&D internationalization to compensate for the lack of domestic technological richness.

Recently, empirical research has provided more direct evidence for the view that different motives for conducting innovative activities abroad influence firms' valuation of locational features. An example is the study by Chung and Alcácer (2002) that, however, is not exclusively focused on FDI in R&D. The authors find that while in general companies targeting the US do not seem to be attracted by the host location's technological resources, MNCs operating in research-intensive industries are. This result points to the heterogeneity of location choices across different FDI investing motives: MNCs that are mainly attracted to the US by the prospect to access a large and sophisticated market are not sensitive to local technological resources, while knowledge-seeking FDI – which are more likely in R&D intensive industries – seek to establish in areas that are endowed with prominent technical capabilities. The authors also find that MNCs from both laggard and leading countries seem to ascribe great importance to local R&D intensity, thus suggesting that knowledge-seeking FDI could either serve as a mean to catch-up in areas of relative technological weakness, or be used to complement existing technological advantages.

Subsequent research also demonstrates that MNCs that invest abroad to conduct basic and applied research are more likely to establish their affiliates in host countries featuring a strong endowment with technological resources, while the location of units that are responsible only for design and development tasks is merely driven by the host country market size (Shimizutani and Todo, 2008).

More recently, scholarly attention has leveraged the idea of a "spiky" world, characterized by a highly uneven distribution of technological resources worldwide, to suggest that contemporary R&D locational trends are increasingly driven by the quest for technological excellence (Castellani et al., 2013). According to this view, MNCs headquarters prefer to establish foreign R&D activities in the most sophisticated technological clusters worldwide, regardless of the geographical distance that separates such locations from the firm home-country. Such perspective is consistent with previously documented rationalization dynamics according to which, after a period of generalized foreign R&D investment, MNC headquarters seek to streamline their foreign innovative activities to privilege those that can more powerfully contribute to firms' technological evolution (Gerybadze and Reger, 1999). In this regard, three types of locations can be identified (Gerybadze and Reger, 1999):

- *leading-edge locations*, where the most complex and sophisticated knowledge in a specific technological field is generated;
- *advanced locations*, by and large overlapping with those countries in the developed world that are characterized by a dynamic market and a superior scientific infrastructure;
- *less developed locations*, where both the demand and the local technological capabilities are relatively backward.

Based on such distinction, it can be argued that MNC headquarters will strive to identify the world-class centers of technological excellence (i.e., leading-edge locations) to locate R&D units that will be responsible to perform the most complex and strategic innovative activities of the group, while selecting the other two location types for less critical tasks (Gerybadze and Reger, 1999).

Beyond the focus on home and host technological features, literature on R&D foreign location choices has documented a variety of additional determinants, such as the existence of large host country markets (Fors, 1996; Håkanson, 1992; Kumar, 1996, 2001; Zejan, 1990), the availability of low-cost R&D human resources (Kumar, 2001), the psychic distance among the home- and the host country (Håkanson, 1992).

Interestingly enough, policy factors, such as trade-regimes and the degree of intellectual property right (IPR) protection, have not been found to play a univocal influence on R&D location choices. Strong IPR regimes in particular seem to attract FDI in R&D in developed countries, while they do not significantly affect R&D location choices in emerging countries (Kumar, 1996). It is possible to speculate that, while MNC headquarters tend to be more strongly concerned about the protection of their research outputs when performing new and high-potential innovative activities, which are usually conducted in advanced contexts (Thursby and Thursby, 2006), knowledge appropriation dangers are perceived as less critical in emerging countries where the most important locational attractors are the high-growth potential of the local markets and the endowment with cheap R&D workforce, and where MNCs usually conduct less strategic innovation projects.

On the whole, this review highlights two main limitations: a strong focus on technology-related aspects that leaves other critical influencing factors, such as those relating to institutional and social features, largely unexplored, and an almost complete emphasis on the country as the relevant geographical scale of analysis, which hinders a more fine-grained account of sub-national sources of heterogeneity. The latter aspect in particular is widely discussed in Chapter 4.

More in general, it can be argued that the need to adjust products and processes to the specificities of international markets to exploit existing MNC advantages has been and will likely be the dominant motivation for the establishment of foreign R&D. Despite this trend, the existence of foreign subsidiaries involved in more *active* technological roles provides a valid argument in support of the idea that R&D internationalization, particularly in its qualitative meaning, is in fact a noteworthy trend, especially given the inherent complexities that characterize creative R&D activities, as opposed to more standardized tasks.

Adaptive R&D is more geographically widespread and quantitatively significant, because it is expected to support foreign production and commercialization functions that are often highly globalized. Conversely, more sophisticated forms of R&D are likely to be performed in a restricted selection of geographical areas that ensure the access to particularly attractive combinations of locational advantages. Hence, while market-driven innovative activities tend to be more pervasive, both in geographical and in quantitative terms, technology-driven foreign investment can be expected to be much more "irregular".

The heterogeneous quantitative patterns of adaptive and creative R&D activities should not advise toward the interpretation of the latter as a trivial phenomenon, as this heterogeneity can be explained by accounting for the inherently different nature and motivations of the two types of FDI in R&D. On the contrary, the very existence of technology-driven, creative foreign activities, which reveals MNCs' pursuit of geographically dispersed technological opportunities, is a sufficient reason to investigate the wide-ranging facets of this phenomenon.

2.4 Organizational challenges in multinational innovation

Ideally, decentralized foreign R&D subsidiaries should be able to perform a wide range of tasks in order to effectively contribute to the process of knowledge generation and accumulation that is critical for the continuous renewal of MNCs' competitive advantage.

First, they should be able to tap into localized pockets of knowledge and expertise, exploit these assets if they are useful to feed their subsidiary-specific innovation processes, but also share them for their potential use within other parts of the MNC organization (Rabbiosi, 2011; Rabbiosi and Santangelo, 2013). The knowledge foreign R&D subsidiaries circulate within the MNC organization often incorporates both inputs acquired externally, that is, in the host location, and new knowledge created internally by the MNC units. It could also happen that the knowledge sourced locally is not useful for the individual foreign subsidiary, while being valuable for its sister units located elsewhere (Lahiri, 2010). In any case, the internal sharing of local knowledge inputs may uncover several application opportunities in a wide range of distant units.

R&D subsidiaries should also learn to be responsive to innovative ideas that other sister units might share within the MNC network, so as to capture potential opportunities to adapt such knowledge inputs to the specificities of their own local context (von Zedtwitz and Gassmann, 2002). This is possible as subsidiaries become familiar with the environment in which they are established by means of informal linkages with competitors, suppliers and distributors, which allows gaining information on local habits and user needs (Zanfei, 2000). When these conditions are met, foreign R&D subsidiaries may act as "interfaces" between generic knowledge that is being shared in the internal network, and context-specific knowledge that is being sourced locally through the development of external networks (Zanfei, 2000).

Finally, subsidiaries should actively seek for solutions to their technical problems inside the MNC organizational boundaries, before searching outside. On one hand, it is possible that units located elsewhere have already faced the same situation. On the other hand, as geographically dispersed units have access to differentiated local knowledge bases, the chances that the solution to the subsidiary's problem is available in one of these affiliates can be expected to be high. Failing to locate already existing knowledge that resides within the MNC organizational network generates a duplication of R&D efforts that creates inefficiencies.

From an organizational viewpoint, the tasks MNCs need to perform to take full advantage of R&D geographical dispersal pose several challenges. For example, the "internal sharing" step is potentially critical, as localized knowledge, which is typically context-specific, needs to undergo a non-trivial process of *generalization* in order to be effectively transferred to other MNC units and utilized in environments other than those in which it has originally developed (Zanfei, 2000).

Similarly, for subsidiaries to effectively acquire the technological knowledge developed by their sister units, the sender and the receiver need to be motivated respectively by a willingness to transfer and a willingness to adopt the knowledge to be shared (Gupta and Govindarajan, 2000). Yet, while the sending subsidiary's motivation could be compromised by the prospect to lose the exclusive possession of its locally developed knowledge, the receiving subsidiary's motivation to internalize such knowledge inflows could be affected by "not-invented-here" barriers.

Moreover, it is possible that the sending subsidiary does not recognize the value of locally acquired technology to be shared with sister units, for instance because it does not have an immediate application in its current area of activity. In such case, it will likely have a limited motivation to understand completely the structure of knowledge; as a result, its transfer capability could be undermined (Lahiri, 2010). Communicating knowledge that is not fully understood is difficult, and it is so especially when the sender and the receiver are located at a distance, which reduces the richness of the communication channels.

Also the "search" task is not an easy one: while a greater number of geographically distributed R&D labs increases the likelihood to find the appropriate knowledge, it also raises the search costs the subsidiary has to incur to locate it (Lahiri, 2010).

In other words, R&D decentralization, and its subsequent geographic dispersal, is a double-edged sword. It increases the knowledge sourcing opportunities, ensures the convenient access to innovation-specific resources, widens the breadth of technological search and places the MNC in a privileged position to be constantly up-to-date about the most promising technological frontier advances, regardless of where they happen. Simultaneously, it raises the R&D organizational complexity, which encompasses a wide range of costs.

It has been suggested that, due to the combination of such benefits and drawbacks, R&D geographical dispersion exerts a curvilinear effect on the overall MNC innovation performance (Lahiri, 2010). With increasing levels of R&D dispersal, the advantages of a geographically distributed access to knowledge-based resources outweigh the costs of managing knowledge across the different R&D sites. However, after a certain degree of international distribution of R&D activities, coordinating innovative units located worldwide becomes too complex as several organizational tensions emerge, and the resulting inefficiencies exceed the benefits of gaining access to additional knowledge sources from new localities. Accordingly, literature has suggested that the global distribution of R&D activities has often generated exaggeratedly complex organizational architectures (Gerybadze and Reger, 1999).

2.4.1 Orchestrating geographically dispersed innovation activities

MNC headquarters have a great responsibility as their orchestration of R&D dispersed activities could both augment and reduce the value created by foreign subsidiaries. Yet, orchestration is not an easy task, as it comes with several organizational tensions.

The most evident organizational tension associated with the geographic dispersion of R&D relates to the trade-off between autonomy and control (Florida, 1997). Control is defined by the set of mechanisms that headquarters leverage to regulate subsidiary activities, combining elements of centralization (when all decisions are taken centrally by parent companies), formalization (when decisions are taken based on established procedures) and socialization (when decisions are inspired by organizational expectations and shared values) (Nobel and Birkinshaw, 1998).

Autonomy can be considered as the most critical facet of control, in that it mirrors the extent to which central offices have the authority and power to affect decentralized R&D decision-making processes. Exploiting the opportunities embedded in overseas R&D sites requires subsidiaries a certain level of autonomy to unleash creativity and ease the generation of new ideas. Yet, headquarters need to maintain an appropriate level of control and coordination of their foreign R&D affiliates to preserve the internal coherence of the MNC innovation strategy.

The tension between autonomy and control is effectively explained in the light of the resource-dependence relationships that link headquarters to subsidiaries (Pfeffer and Salancik, 1978). Originally, subsidiaries were completely dependent on headquarters' centrally developed technology to perform their local activities. With the emergence of geographically distributed clusters of relevant technological resources, R&D units have gained access to distinctive pockets of knowledge and competences, useful to feed their innovation processes. As the reliance on the parent company's technological portfolio gradually diminishes, the power relationship between subsidiaries and headquarters shifts toward new balance points.

For instance, the literature on reverse knowledge transfer, looking at the headquarters-subsidiary dyadic relationships, shows that the individual subsidiary may become an extremely important knowledge source for the parent company. Accordingly, the design of the different coordination mechanisms deployed within the MNC organization is not independent from the subsidiary's role and degree of autonomy (Rabbiosi, 2011).

In other words, while headquarters used to enjoy the most central position in the MNC network, with subsidiaries maintaining very few external linkages with agents in their host location, R&D decentralization offers to foreign units the opportunity to become more and more integrated within their local network. Through local participation and the acquisition of "insider" positions, subsidiaries can source strategic resources locally. Leveraging high-quality linkages with local partners may therefore contribute to grant subsidiaries a higher bargaining power in the face of the parent company. However, the more subsidiaries tighten their linkages with the local network, the higher the risk that they will divert from corporate objectives, thus reducing the degree of MNC internal cohesion (Asakawa, 2001).

The design of control and coordination mechanisms also relates to another organizational aspect that has to be carefully managed, namely communication. In the context of R&D, communication covers the wide-ranging information flows involving managers in different R&D labs, which can be broadly classified into (1) vertical communication flows, which connect R&D units with the head-office, thus often acting also as control tools; (2) lateral communication flows between R&D sister units or between R&D units and other functions, which are critical to foster the interaction among interdependent affiliates; (3) communication flows with external actors, such as suppliers, customer or local universities, which ensure local responsiveness by enabling increasing integration in the local context (Nobel and Birkinshaw, 1998).

Several issues arise in the management of communication flows as R&D becomes geographically dispersed. On one hand, R&D geographical distribution generates potential risks for inter-unit communication, resulting not only from geographic distance but also from cultural barriers. On the other hand, as already suggested, extensive external communication has both benefits and drawbacks. While it grants a privileged access to local strategic technological resources, it may also gradually isolate foreign units from the MNC internal network. It follows that communication lines within a MNC internal network have to be effectively configured, so as to ensure a simultaneous attention to internal cohesion, local responsiveness and efficiency in terms of exploitation of potential synergies.

On the whole, the foregoing discussion suggests that both coordination and autonomy encompass costs. For instance, tight coordination generates transaction costs, opportunity costs arising from a lack of sensitiveness to local markets and resources, as well as the costs of developing and maintaining an appropriate information system (Gassmann and von Zedtwitz, 1999). On the other hand, high subsidiary autonomy generates duplication costs, agency costs arising from conflicting objectives, efficiency costs due to suboptimal R&D size at the different foreign sites, as well as opportunity costs arising from the inability to exploit potential synergies across locations (Gassmann and von Zedtwitz, 1999).

Beyond these costs, R&D geographic dispersion entails other organizational dilemmas that are strongly related to the tension between autonomy and control: the juxtaposition between process and hierarchy, creativity and disciplines, short-term and long-term orientations points to the wide range of trade-offs emerging from the search for the optimal management of R&D organization within MNCs (von Zedtwitz et al., 2004).

With increasing R&D geographic dispersion, the hypothesized value-adding role of headquarters in subsidiary innovation could also turn into value-destruction. While traditional *rational* views assume that parent companies are able to discern whether they possess the relevant knowledge to intervene in subsidiary local innovation processes, more recent studies inspired by the "sheer ignorance" perspective suggest that, in certain situations, headquarters' ability to recognize the knowledge that is required to positively contribute to a subsidiary's innovative activities may be limited (Ciabuschi et al., 2011). In such case, normative expectations may drive the headquarters to intervene into the subsidiary's innovation activity despite its lack of appropriate knowledge (Ciabuschi et al., 2011). As the level of R&D geographic spread increases, a higher number of subsidiaries will gradually integrate within their local network and engage with context-specific technological knowledge that parent companies are less likely to be familiar with. As a result, headquarters' involvement into subsidiary innovation will more likely be detrimental, rather than beneficial, with obvious consequences on subsidiary innovation performance.

Though the foregoing discussion reveals manifold potential risks, research demonstrates that MNC headquarters are not powerless in the face of the organizational challenges posed by R&D geographic dispersion. Inspired by network logics, they have increasingly learned to develop effective organizational structures that utilize newer control and coordination mechanisms based on shared language, values, behavioral norms and integrated communication (Hedlund, 1986). Scholars have labeled such innovative organizational models differently. For instance, Bartlett and Ghoshal (1990) identified the departures from the traditional centralized MNC structure with the so-called "transnational" model, while Hedlund (1986) proposed the "heterarchical" MNC model.

Despite the diverse labels, the resulting conceptualization converged toward an organizational structure alternative to hierarchy, which is able to capture the increased complexity of activities and relationships within and outside the MNC, and in which resources and competences are distributed, rather centralized, throughout the organization, control is accomplished through normative integration, and relationships are not only vertical but also lateral (Birkinshaw and Morrison, 1995). In other words, moving from hierarchy to heterarchy, some MNCs have developed a multifaceted network of relationships, and identified more flexible and effective decision-making processes and integration mechanisms that enable to exploit advantages arising from the global distribution of the firm activities (Hedlund, 1986).

Although it is not possible to state that this evolution has been driven solely by the need to adapt to the requirements of an increasing geographical distribution of technological resources, the new organizational arrangements seem to have granted MNCs an exclusive advantage in the complex management of the geographical boundaries of knowledge sharing, particularly when control and coordination mechanisms are effectively adapted to the specific conditions of each subsidiary (Ghoshal, 1986).

The "spiky world" view suggests that specialized technological knowledge tends to accumulate in specific locations in a path-dependent way, such that only selected places across the world ensure the access to the "best" knowledge in a given technological field (Castellani et al., 2013). Because firms are not indifferent among different locational alternatives for their R&D activities, they seek to maximize their foreign investment in R&D by embedding in world-class technological clusters, no matter how distant they are from their home-country. To be able to manage innovative facilities regardless of their "remoteness", MNCs – or at least some of them – have developed efficient mechanisms and organizational procedures to codify, process and circulate knowledge across geographical space (Castellani et al., 2013).

Overall, the advantages associated with the geographic dispersal of R&D activities are likely to be fully achieved only by those MNCs that may leverage an appropriate inter-organizational network. This statement is at least partly supported by empirical evidence showing that while some MNCs continuously push toward R&D internationalization, in other firms, early geographical dispersal has been followed by a "re-centralization" effort, aimed at reducing the distribution of R&D activities within a limited number of locations that ensure an easier coordination (Gerybadze and Reger, 1999).

Effective network models overcome the traditional hierarchical structures based on the unidirectional transfer of technological knowledge from the parent company to foreign subsidiaries, and foster the emergence of new organizational paradigms in which geographically distributed MNC units not only passively absorb knowledge produced elsewhere but also actively create and circulate new knowledge, generated through the leverage of local sources of information and skills.

There exist several classifications of the organizational solutions for managing R&D in MNCs, which have been updated along the years to account for the evolution of management styles and procedures.

For instance, merging Bartlett's (1986) distinction between international, multinational, global and transnational corporations with Perlmutter's (1969) behavioral orientation model, Gassmann and von Zedtwitz (1999) propose five forms of R&D organization, suggesting that MNCs' organization of R&D subsidiaries evolve in the search for the ideal balance between control and coordination:

1. the *ethnocentric centralized R&D organization*, which concentrates all innovative activities at home to keep the highest control and coordination;

2 the *geocentric centralized R&D organization*, which keeps R&D at the center to retain the benefits of physical concentration, while fostering the sensitivity to foreign markets through the leverage of employee international mobility and international recruiting policies;
3 the *polycentric decentralized R&D organization*, that is based on a decentralized configuration of highly autonomous R&D affiliates that operate without the supervision of the corporate R&D structure;
4 the *R&D hub organization*, which works through a node structure that helps presiding relevant technological fields, but ascribes a strong authority to the R&D center that performs the most advanced innovative activities in the home-country;
5 the *integrated R&D network organization*, that fully overcomes the traditional dominance of the home-base R&D center, which is only one among many integrated and interdependent units, each playing a strategic role based on specific competences, and pooling their technological efforts for the benefit of the entire company.

To account for the differences between science-based and engineering-based tasks, which could be heterogeneously affected by centrifugal and centripetal forces, von Zedtwitz and Gassmann (2002) develop the following taxonomy and suggest that MNCs move across the four proposed archetypes depending on the nature of their R&D internationalization drivers:

1 *national treasury R&D*, characterized by the home-country concentration of both research and development activities;
2 *technology-driven R&D*, consisting of geographically distributed research but domestic development;
3 *market-driven R&D*, in which research is performed domestically and development is realized in different foreign locations;
4 *global R&D*, characterized by the geographic dispersal of both research and development.

With no claim of exhaustiveness, this brief review offers a general perspective on the potential arrangements for managing R&D across space. Moreover, along with many of the other works surveyed in this chapter, it contributes to emphasize the importance of adopting a dynamic approach to the interpretation of R&D internationalization-related phenomena. This view is not new to the IB literature, and rather finds its origins in the prediction of Ronstadt (1978) who, in a seminal paper on the technological activities in MNCs, hinted at the evolutionary dynamics by which initial R&D assignments in foreign subsidiaries trigger the development of substantial capabilities, ultimately leading to product innovations that target foreign or even global markets.

Following this idea, both the literature on the nature of foreign innovative activities and the research stream on the organization of international R&D have stressed that R&D tasks and the architecture of innovative activities in MNCs are far from being immobile, and instead change over time depending on the trends involving foreign markets, foreign clusters of science and technology, and the interaction dynamics between these and foreign subsidiaries.

As the next chapters show, this evolutionary view is a major aspect that should be accounted for to understand the changing role of subsidiaries in the current IB environment, as well as the critical effect host locations exert on this process.

Part II
A Multilevel Approach to the Study of Geographically Dispersed Innovation in Multinational Firms

3
Perspectives on Subsidiaries

Abstract: *This chapter shifts the focus of the analysis of MNC innovation from headquarters to subsidiaries. It reviews IB literature on subsidiary evolution, and discusses the internal and external conditions leading subsidiaries to gain and further develop active roles in the management of local innovative activities. Leveraging innovation management literature, it describes subsidiaries as strategizing actors that govern their local knowledge assets in the pursuit of their subsidiary-specific objectives and incentives. Specifically, it suggests that subsidiaries' local innovation management is driven by a combination of knowledge creation and knowledge protection imperatives. In the quest for such objectives, subsidiaries are able to actively and dynamically adapt their knowledge strategies to changing internal and external conditions.*

Keywords: active subsidiaries; internal and external networks; knowledge creation; knowledge protection

Perri, Alessandra. *Innovation and the Multinational Firm: Perspectives on Foreign Subsidiaries and Host Locations*. Basingstoke: Palgrave Macmillan, 2015. DOI: 10.1057/9781137555441.0010.

3.1 The evolution of subsidiary-level research

So far, this volume has examined research and development (R&D) internationalization by focusing on macroeconomic surveys on the changing dynamics of this phenomenon and on the evolution of related managerial aspects from the multinational corporation (MNC) headquarters' perspective. It has been highlighted how the new geography of technological innovation has progressively triggered a significant change in the way foreign R&D activities are planned, implemented and subsequently coordinated in MNCs. In this mutable scenario, for several years, scholarly emphasis has converged on MNCs' headquarters, given that these actors have the power to take decisions on foreign direct investment (FDI) in R&D.

Corporate headquarters certainly play a critical role in the management of the firm's international network of R&D units. Accordingly, the previous chapter has analyzed the wide range of motivations that may drive headquarters to internationalize R&D and the resulting factors influencing location choices; it has reviewed the different tasks parent companies may assign to foreign labs, and discussed the complex organizational challenges they have to manage to effectively orchestrate the network of foreign R&D units.

To complement such headquarters-centered perspectives on R&D internationalization, this chapter aims at emphasizing the active role subsidiaries may play in the management of the MNC technological resources. This has long been an overlooked aspect in both the academic research and the managerial practice. On one hand, traditional theoretical models of MNC innovation identified the corporate headquarters as the sole engine of technological development; in such models, subsidiaries served as passive implementers of a higher-level technological strategy. On the other hand, the reality of the leading enterprises has often revealed a certain degree of resistance to subsidiary-level strategic proposals and initiatives, which used to be perceived as subversive actions, and interpreted as dangerous deviations from the firm's centrally planned strategic path and corporate objectives.

What changed this picture? A first factor that has triggered the debate on subsidiaries' role in MNC innovation management is the attention that has gradually been ascribed to subsidiary-level strategy-making activity and, accordingly, to its contribution to the firm value creation. While it is undeniable that corporate headquarters play a prominent role in the MNC decision-making process and in the resulting build-up of firm-specific competitive advantages, during the 1990s, international business (IB) literature has started to point to the increasing involvement of subsidiaries in the process of formulation and implementation of the MNC strategy. This trend is clearly very general and may involve any subsidiary, regardless of their area of competence; yet, when high value-added activities – such as R&D and innovation – are involved, it may be expected to spawn more pervasive consequences for the entire MNC's strategy and performance.

Moreover, scholars have progressively realized that subsidiary roles, as originally assigned by the head-office, are not immutable, but rather may change depending on both the subsidiary local activities and the opportunities and challenges embedded in the host location. Again, while they apply to a variety of subsidiaries and roles, these dynamics are particularly critical when R&D units are concerned. In fact, because the technological resources residing in host locations are of fundamental importance for subsidiaries' innovation processes, and locations are in continuous evolution, the exploration of how subsidiary innovation behavior varies depending on its interaction dynamics with the local context is critical not only to understand subsidiary-level innovation, but also to fully appreciate the expected technological contribution foreign subsidiaries may offer to the MNC as a whole.

The rapid pace with which new locations from emerging countries have gained relevant positions in the geography of innovation has also contributed to attach a growing relevance to foreign subsidiaries' innovation strategy, from both an academic and a managerial perspective. To preside the vibrant dynamics of these contexts, which increasingly act as sources of simple, yet path-breaking ideas leading to novel products, services and modes of doing business, MNCs have realized the importance of "being there" through their own R&D investment.

This demands foreign subsidiaries not only to understand entirely different contexts but, most importantly, to engage in local entrepreneurial activities that could enable them to exploit business opportunities embedded in these areas in a more effective and timely way. Simultaneously, it exposes parent companies to the need to delegate higher R&D responsibilities to units located in these contexts, in the recognition that the centralized management of the full range of a firm's globally distributed activities from a single location becomes too difficult when prominent differences exist between regions, as it happens for MNCs operating in both traditional and emerging markets (Mudambi, 2011).

More in general, headquarters and central R&D managers have come to realize that in order to ensure an effective management of the innovation-related opportunities and threats that may arise from distant locations, foreign subsidiaries need to be endowed with a higher degree of autonomy in strategy making. In fact, technological processes become increasingly rapid thus prompting the emergence of unexpected events and conditions that have to be managed in a timely manner. The need for quick decision-making capabilities in innovation-related issues calls for a greater involvement of agents that are "closer" to where such events and conditions occur, thereby shifting the attention on the role played by foreign subsidiaries.

The antecedents and consequences of these modifications in the structure and strategic management of innovation in MNCs are analyzed more thoroughly in this chapter.

3.2 Changing roles: from passive to active subsidiaries

Most early research on MNC subsidiaries tends to overlook the topic of strategy (Birkinshaw and Morrison, 1995). Though strategy-making at the headquarters-level has immediately attracted scholars' attention, given these actors' fundamental ability in managing inherently complex market and business portfolios and organizational structures, subsidiary research originally focused on the relationships linking these actors to headquarters, and only subsequently started to analyze the characteristics of the local environment where they operate. However, these studies generated ambiguous results, which some scholars ascribed to the lack of a proper account of the different strategic roles subsidiaries may play (Birkinshaw and Morrison, 1995; Gates and Egelhoff, 1986).

It is probably with the emergence of the global strategy literature (Bartlett, 1979; Prahalad and Doz, 1981) that the issue of subsidiary strategy entered the stage of IB research. Scholars have proposed several classifications of subsidiaries' strategic roles, which have been effectively integrated in Birkinshaw and Morrison's (1995) typology.

The authors distinguish between the *local implementer*, the *specialized contributor* and the *world mandate* subsidiary.

The *local implementer* almost overlaps with the subsidiary type that White and Poynter (1984) named the "miniature replica", as well as with Jarillo and Martinez's (1990) "autonomous" affiliate. This subsidiary is responsible for a limited geographic area and its product scope is narrow. It creates value by tailoring the MNC's global products to the specificities of the local demand or business practices, thereby being very widespread in firms adopting multidomestic strategies (Porter, 1986).

Moving toward higher value-creating roles, the *specialized contributor* is a subsidiary whose tasks are very focused, as they relate to specific functions in which it possesses significant competences and expertise. Its activities are thus highly interdependent with those of other sister units, as synergies may exist among the respective tasks. This subsidiary type resembles the "receptive" affiliate described in the work of Jarillo and Martinez (1990), and may be though to cover both the "rationalized manufacturer" and the "product specialist" categories of White and Poynter's (1984) classification, depending on the breadth of the value-adding activities and products it is tasked with.

Finally, the *world mandate* subsidiary, analogous to the "active" subsidiary type proposed in Jarillo and Martinez's (1990) classification, is normally responsible for a product line, a business or an element of a business, considering the skills it has developed in such area. In terms of geographic scope, its responsibility is either global or regional. Hence, while the activities relating to the specific product line or business are integrated worldwide, they are orchestrated by the subsidiary, rather than by the headquarters. It follows that such subsidiaries are rather autonomous, as their product development processes are managed independently from the central office.

There are clearly very significant differences across the subsidiary roles described above. The recognition of such subsidiary-level heterogeneity marked a switching point in IB research. In fact, by acknowledging that not all subsidiaries are the same in terms of geographic scope, autonomy, breadth of value-adding activities, capabilities and expertise, scholars formalized the idea that they can heterogeneously contribute to MNC performance.

Despite this critical step, it took several years for scholars to explicitly adopt an approach that considers strategy-making at the level of subsidiaries as worthy as strategy-making at the level of the headquarters or of the corporate group as a whole, as a research topic.

Not surprisingly, almost the full range of studies focusing on subsidiary heterogeneity used the term "role", instead of pointing to differences in subsidiary "strategy". As suggested by Birkinshaw and Morrison (1995), there is more than a merely semantic issue behind this choice. In fact, the notion of "role" presumes a centrally assigned responsibility, with which subsidiaries are demanded to comply. Conversely, the idea of "strategy" moves the focus of the analysis on the subsidiary itself and on its individual choices, thereby acknowledging that they could potentially differ from the plans originally designed at the head-office level.

In other words, in order to allow for a *genuinely active* subsidiary model, one should account for the possibility that subsidiary activities do not result exclusively from a deterministic process led by the headquarters, but rather can change depending on variables that are not under the full control of parent companies.

Looking at the reality of contemporary MNCs, it has become increasingly evident that some overseas subsidiaries undertake dynamics involving growth and capability development that cannot be ascribed to headquarters' infusion of centrally accumulated resources. Clearly, there are other variables that influence some subsidiaries' ability to evolve beyond the tasks and responsibilities originally assigned by the head-office.

Several scholars concurrently pointed to the dynamic process that governs subsidiary activities in the host location. Yet, the most established model of subsidiary evolution is probably the one developed by Birkinshaw and Hood (1998). The authors suggest that, beyond the *head-office assignment* of the subsidiary role, which represents both the starting point of the subsidiary evolution process and an important determinant of what the subsidiary will become in the future, two other factors enter this dynamic, namely the *subsidiary choice* and the *local environment determinism*.

At any given point in time, head-office assignment of subsidiary roles depends on subsidiaries' endowment with resources, capabilities and expertise, as well as on the perceived strategic importance of the market in which they operate. If considered in isolation, this view is in line with traditional literature that has conceived subsidiaries as mere instruments in the hands of headquarters: these actors cannot refuse to play a given role, if this emanates from the MNC center, although headquarters may also choose to increase their commitment to a given market over time, particularly if the local subsidiary has meanwhile developed a good fit with its external environment, leading to knowledge accumulation and raising performance.

Although this deterministic approach was consistent with established literature, empirical evidence suggested that it was not sufficient to explain subsidiary evolution, for various reasons. For instance, while head-office assignment was clearly the most critical determinant of subsidiary role in the early stages of this evolutionary process, that is, when subsidiaries tend to be endowed with limited and lower-order capabilities, existing theoretical perspectives could not fully explain the gradual accumulation of high value-adding activities and strategic responsibilities in specific subsidiaries. The missing link lies in the observation that, over time, subsidiaries may become much more than passive implementers, as they often come to develop specialized competences on which other actors in the MNC tend to depend.

To accept this view, one should overcome traditional hierarchical views stemming from Vernon's product lifecycle model (1966), to embrace a network perspective of the MNC, in which resource-dependence matters, and where originally "peripheral" actors may become increasingly central in the power relationships governing MNCs' organization given the distinctive capabilities they accumulate over time (Forsgren et al., 1992). In such views, subsidiary managers may strategically act to demonstrate the value of their acquired capabilities, thereby leading corporate management to assign them an enhanced charter.

Clearly, the process of subsidiary evolution cannot be fully understood if we fail to account for the idea that the conditions in the environment in which organizations operate may strongly constrain and nurture their ability to reach their goals (Hannan and Freeman, 1977). Applied to the context of subsidiaries, this argument gains a striking importance as these units can benefit from the resource-base embedded in the external network of local actors and conditions in fundamental ways. For instance, because this resource-base is likely to be very different from the resource portfolio available at home, it may potentially lead to the development of distinctive subsidiary-specific capabilities.

In other words, the wide-ranging opportunities for knowledge diffusion and creation associated with geographical proximity make the host environment and the knowledge dynamics that occur locally a critical driver of subsidiary capability accumulation. Hence, through the leverage of local resources and relationships, subsidiaries develop unique and specialized skills that may foster autonomy from the internal organization.

On the whole, the model of subsidiary evolution recognizes the importance of subsidiaries' *freedom*, and overcomes the idea that their destiny is strictly and inexorably determined by the headquarters' will (Birkinshaw and Hood, 1998).

It should be noted that the acquisition of an expanded charter by a given subsidiary tends to occur through a dynamic managerial approach, in which the subsidiary entrepreneurial pull plays a fundamental role. Entrepreneurial actions serve as a critical building block for the development of perspectives on subsidiaries as active strategy-makers. In fact, they are undertaken *consciously* and *deliberately* (Bouquet and Birkinshaw, 2008), pointing to a clear strategy-making process. Subsidiaries that are willing to increase their strategic importance within the MNC network dynamically search for new market opportunities, which often arise in their own local environment, and seek to cultivate the resources required to exploit them. The presence of an entrepreneurial leader is therefore important for the emergence of subsidiary initiatives. More generally, the availability of a widespread pool of entrepreneurial human resources certainly helps subsidiaries to upgrade their role, as these employees increase the responsiveness of the unit, as well as its ability to source from a large and diversified range of new ideas (Birkinshaw, 1997). Hence, both subsidiary leadership and a diffused entrepreneurial culture support the accumulation and further enrichment of specialized resources, which in turn appear to make subsidiary initiatives more likely (Birkinshaw et al., 1998).

Along with the subsidiary's entrepreneurial attitude, its past performance, intended as the ability to meet or even exceed the head-office's expectations, is a critical factor that corporate managers account for when they evaluate the possibility to expand the subsidiary's responsibilities (Hood et al., 1994). A positive record of results indeed reduces the uncertainty embedded in such an investment decision.

On the whole, these reflections suggest that subsidiaries do not *happen* to gain a better charter; rather, they invest in a process of *capability and reputation building* that lays the basis for the subsequent attempt to gain headquarters' recognition of their higher-level skills.

It should be noted here that hierarchical corporate managers often perceive the same subsidiary initiatives that could foster the acquisition of improved charters as dangerous manifestations of the subsidiary's independence. It is certainly not infrequent that subsidiary managers use entrepreneurial actions to develop their own "empires" regardless of the interests of the MNC as a whole. Yet, frustrating every subsidiary initiative for fear of opportunistic behaviors means heavily reducing the MNC's chances to exploit ideas and skills residing in distant locations. In fact, assuming that subsidiary managers are in a privileged position to assess the value of their own resources, it should be their obligation to actively seek for the best opportunities to deploy these resources to the advantage of the entire MNC (Birkinshaw et al., 1998).

The idea that subsidiary capabilities may have a positive influence beyond the boundaries of the unit and of the geographic scope of which it is responsible leads to the notion of "center of excellence". Centers of excellence are organizational units within foreign subsidiaries embodying a set of advanced capabilities, which the MNC headquarters formally recognize as a fundamental source of value creation (Frost et al., 2002). In the light of such explicit recognition, centers of excellence are required to deploy and circulate their own capabilities within the rest of the MNC network, so as to enable other sister units, as well as the head-office itself, to take advantage of its superior capabilities. Though the head-office's infusion of intangible assets, such as technical knowledge and expertise, is key to establish as a center of excellence, such higher-order strategic roles could not exist without the inherent ability residing in some subsidiaries' organizational units to proactively orchestrate, leverage and augment their competences.

The existence and importance of centers of excellence within major MNCs endorses an evolutionary view of capability development consistent with that proposed by Birkinshaw and Hood (1998), where MNC subsidiaries – aided or constrained by internal and external conditions – accumulate resources and competences over time in a path-dependent way, which eventually drives them to gain a strategic leadership in specific areas of activities within the broader multinational network (Frost et al., 2002).

The ability of a subsidiary to give rise to a center of excellence to create value for its corporate group offers evidence of the existence of significant subsidiary-level strategy-making and strategy-implementation processes, and accentuates the need to investigate strategy at the level of the individual subsidiary. Such need is formally acknowledged by the strand of literature that separates subsidiaries based on their *competence-creating* versus *competence-exploiting* role (Cantwell and Mudambi, 2005).

Both subsidiary types engage in innovative activities. However, they are very different. Compared to competence-exploiting units, competence-creating subsidiaries perform a higher level of R&D, and their technological posture is more creative. In these organizational units, the rationale for performing local R&D has changed over time, overcoming the focus on their most immediate market (Cantwell and Mudambi, 2005). The existence of these subsidiaries is not independent on the firm profile. In fact, leading MNCs are more likely to place creative R&D activities abroad. In turn, the higher geographical spread of these firms' innovative activities increases the chances to find propitious conditions for technology creation in specific locations, where R&D activities consequently become more significant (Cantwell and Mudambi, 2005).

Through the orchestration of a network combining both competence-exploiting and competence-creating subsidiaries, MNCs simultaneously pursue exploration and exploitation, in a way that closely resembles the duality of organizational objectives proposed by March (1991). As suggested by the organizational learning theory, competence-creating roles presume experimentation capabilities that could promote the development of fresh ideas and the exploration of novel opportunities. This requires not only a certain degree of autonomy, but also an ability to take strategic decisions that is essential to manage complex R&D processes.

This brief overview of the evolution of subsidiary research has revealed the need to portray subsidiaries as potentially advanced actors, able to understand the value of their resources and to define long-term objectives for the development of their role within the MNC internal network.

The emergence of such creative subsidiaries is key to the MNC strategic evolution, not only because of their ability to enrich the resource-base of the firm, but also in the light of a higher cross-subsidiary heterogeneity arising from the diversity of location-specific advantages to which they can access (Cantwell and Mudambi, 2005; Pearce, 1999).

Clearly, not all subsidiaries will grow to become creative or strategically active. Rather, it is reasonable to expect that only a limited number of them will be able to effectively accomplish all the steps that allow for a strategic evolution of their role. Despite the narrow scope of this phenomenon, its potential effect is such that it deserves explicit and full recognition in all the streams of IB research.

3.3 Internal and external drivers of subsidiary evolution

To ensure that we gain comprehensive understanding of subsidiaries' evolution toward strategically active roles, this phenomenon has to be considered in the broader context in which it occurs. In particular, it is essential to pay attention to the role of subsidiaries' internal and external networks, as these act as powerful sources of incentives. This approach is consistent with perspectives that conceive MNCs as a double network structure (Castellani and Zanfei, 2006), according to which two sets of dynamics influence the creation and use of knowledge in contemporary MNCs: on one hand, the wide-ranging relationships between headquarters and subsidiaries as well as among sister units, and on the other hand the linkages with local external actors, such as suppliers, customers and other institutions.

3.3.1 The influence of the internal network

Although the MNC's internal network has been analyzed from a variety of viewpoints, consistent with the objectives of this work, we shall focus on its role in influencing subsidiaries' technological activities. In this regard, literature has offered some insights on the relationships linking the MNC corporate culture, structure, organization and internal interaction patterns to the transition of subsidiary technological behavior from passive to active. In other words, starting from the very early work mapping the evolution of the MNC innovation processes from "local-for-local" to "local-for-global" (Bartlett and Ghoshal, 1990), this literature covers all those studies that have investigated how factors relating to the structure and properties of the MNC internal network could influence the subsidiary technological posture.

In the traditional theory of the MNC, internal relationships are dyadic and hierarchical; they usually emanate from the parent company, which take decisions based on ownership rights as well as on its disproportionate endowment – relative to subsidiaries – with valuable resources that constitute the basis for the firm's ownership advantage.

As intangible resources become more dispersed within the MNC internal network, MNCs evolve from *technology creators* to *technology organizers* (Cantwell, 2001). Hence, from being the only source of technology to which subsidiaries can hope to have access, headquarters become the orchestrators of geographically distributed technological resources, some of which may be autonomously generated by individual subsidiaries. As such, their role has been analyzed both as a potential constraint and as an enabler of subsidiaries' technological creativity.

As often highlighted in the foregoing discussion, principal-agent perspectives suggest that subsidiaries' creative initiatives may be perceived as dangerous, as they usually depart from the mandate assigned by the head-office (Birkinshaw, 2000; Birkinshaw et al., 1998). Parent companies may choose to impede the pursuit of entrepreneurial actions, thus limiting subsidiaries' technological creativity. However, headquarters' reaction to subsidiary initiatives could vary depending on the firm corporate culture, on the nature of their strategic leadership as well as on the overall firm responsiveness to novel opportunities.

As a case in point, most studies suggest that corporate integration and control have a negative impact on subsidiaries' local creative activities (Andersson et al., 2005; Jarillo and Martinez, 1990) or, similarly, that increasing decentralization of decision-making power on knowledge and innovation-related issues fosters a subsidiary's active management of its technological agenda (Florida, 1997).

In fact, because high control is often coupled with a willingness to avoid duplication and to maintain a strong internal consistency among the firm's dispersed innovative activities, it is within MNCs with decentralized decision making that the emergence of subsidiary initiatives and the upgrade of subsidiaries' mandates are more likely to happen (Birkinshaw and Hood, 1998). In this regard, the nature of subsidiaries' creative specialization may also be expected to play a role: some subsidiaries could develop radically different competences, compared with the technological trajectory of the firm as a whole, thereby challenging the coherence of the MNC's technological portfolio (Pearce, 1999). In this case, an overly permissive approach on the part of headquarters may be risky, although the underlying fundamentally novel technological opportunities could conceal highly promising strategic paths.

Headquarters also have the delicate task to allocate resources for the development of subsidiary-level projects and capabilities that could contribute to the MNC as a whole, which generates internal competition. Internal competition is expected to play a positive influence on subsidiaries' ability to undertake creative technological behaviors, as it motivates sister units to strengthen their resource and capability base (Birkinshaw and Hood, 1998).

Yet, in the process of evaluating subsidiary capabilities and their technological potential, headquarters could make mistakes, for instance denying sufficient resource availability to subsidiaries whose skills have been underestimated. Subsidiaries are not powerless in the face of such decisions though, as they can influence the central allocation choices. For instance, they can seek to gain attention from corporate headquarters, and have more chances to succeed in this task if they are endowed with upstream capabilities, such as innovation capabilities (Bouquet and Birkinshaw, 2008). It is not surprising that subsidiaries that embody technological knowledge that is critical for the firm as a whole have a higher bargaining ability in their relationships with headquarters (Mudambi and Navarra, 2004). Consistent with this idea, it has been demonstrated that the possession of technology-related power grants subsidiaries the opportunity to exert an influence on the firm's strategic decisions (Mudambi et al., 2014).

At the relationship-level, the quality of the interaction between headquarters and subsidiaries may also favor the subsidiary evolution and, in turn, its local innovative activity (Birkinshaw and Hood, 1998). In fact, subsidiary managers enjoying informal ties with influential decision-makers in the head-office may more easily obtain a positive evaluation of their unit's prospective plans.

Along with headquarters-subsidiary relationships, lateral linkages connecting geographically distributed sister units may also play a role. It has been argued that strong internal linkages with other MNC subsidiaries may act as conduits for the emergence of new ideas, thereby enhancing the subsidiary capabilities. In turn, as subsidiaries develop advanced capabilities and become comparatively more efficient in performing specific tasks that can be executed also to the benefit of sister units, they gain importance within the corporate network and, eventually, assume center of excellence roles (Frost et al., 2002).

As the availability of complementary assets is essential to the full exploitation of a given innovation (Teece, 1986), the MNC's internal network may allow subsidiaries to rely on each other's competences, to collaborate and share knowledge and expertise for their innovative processes, as well as to combine and integrate each other's specialized resources.

Yet, knowledge flows between MNC subsidiaries do not happen automatically. Rather, they depend on the source's and the receiver's respective willingness to share and absorb knowledge, as well as on the richness of existing transmission channels (Gupta and Govindarajan, 2000), which in turn can be influenced by various factors, ranging from subsidiaries' absorptive capacity and cultural differences to the availability of organizational arrangements for inter-unit exchange and socialization.

3.3.2   The influence of the external network

The local context in which subsidiaries operate has been recognized as a critical source of stimuli that may profoundly influence their evolution and, in turn, their technological behavior. Starting from the recognition that host locations do not necessarily act as constraints on MNCs' development, but rather may embody technological opportunities and resources whose value reside especially in their diversity (Hedlund, 1986), a wide literature has developed to investigate this issue.

The quality of the location advantages, intended for instance as the presence of outstanding industrial clusters resulting in agglomeration benefits, positively influences the level of competences a subsidiary may attain. Knowledge-intensive locations may offer critical inputs to feed subsidiaries' innovation processes, thereby motivating them to adopt more active approaches in the management of their technological resources. In this respect, empirical literature has demonstrated that a vibrant local business environment, where the relationships with complementary and competing firms act as fecund sources of information, ideas and opportunities for technological upgrading, may drive subsidiary evolution (Birkinshaw and Hood, 1998).

Interestingly, it has also been shown that, for R&D centers, local environmental features and the inter-organizational relationships with local customers and suppliers are more important than internal factors in explaining the acquisition of center of excellence mandates, whereas the opposite is true for manufacturing affiliates (Frost et al., 2002). The reference of this research stream is Porter's (1990) diamond model that defines the value of locations based on a combination of elements, such as the availability and quality of factor pools, the demand conditions, the presence of related and supporting industries, and the characteristics of local competition.

Local business networks offer platforms for interaction between knowledge sources and knowledge recipients, thereby facilitating the processes of search and transfer (Hansen, 1999; Uzzi, 1997). In fact, networking acts as a social communication process that promotes knowledge sharing and allows for knowledge negotiation among network participants, thereby instigating mechanisms that are crucial to facilitate those social, political and cognitive dynamics through which ideas are designed, communicated and implemented over time.

Building on network theory, literature has more thoroughly investigated the subsidiary's local linkages, particularly when they involve vertical partners. Subsidiary relationships with agents participating to their local business network have been indicated as the channels through which they can attain a better understanding of the local markets and access location-specific pools of competences and resources (Chen and Chen, 1998; Giroud and Scott-Kennel, 2009).

Linkages can evolve in very heterogeneous ways: some may last and become more solid over time, thereby allowing for potential mutual learning, while other linkages are destined to just remain arms-length relationships, much less important for sustaining subsidiary development. To account for such differences, scholars have devoted special attention to the "quality", rather than the quantity of local linkages (Giroud and Scott-Kennel, 2009; Scott-Kennel, 2007), which points to the inherent interactive nature, trust and long-term orientation of selected local relationships. In fact, the social dimension characterizing quality linkages is the major premise for valuable learning, as it allows for effective knowledge sharing.

The investigation into the effect of the relationships with local partners has prompted the development of a research stream on subsidiary embeddedness. Business embeddedness, in the form of the wide-range of direct ties between a subsidiary and its local partners, allows for information exchange on local business matters, and enables subsidiaries to overcome their status of "outsiders". In turn, it instigates technical embeddedness, which comes with the development of interdependence and adaptation on technical issues with local partners (Andersson et al., 2002).

Through the continuous interaction with their local vertical partners, subsidiaries develop efficient routines and specialized technological competences, which can be redeployed to the benefits of sister units (Andersson et al., 2002). In such cases, subsidiaries attain an active role as capability creators for the MNC as a whole, which may also have an impact on their importance within the organization. In other words, when they allow for the development of capabilities that are shared within the internal organization, embedded relationships with the local business network offer to subsidiaries a power base that they can use to influence MNCs' strategic decisions (Andersson et al., 2007).

To develop such established and powerful relationships with local actors, subsidiaries must be endowed with a sufficient degree of autonomy, which may be expected to play its already documented innovation-enhancing effect also through this channel. Research on the *branch-plant syndrome* (Hood and Young, 1976) supports this view, suggesting that when lacking managerial authority subsidiaries develop limited local linkages and tend to remain technologically dependent from the parent company. Accordingly, several authors have suggested that a trade-off exists between subsidiaries' internal and external networking (Castellani and Zanfei, 2006).

3.4 An innovation management perspective

As widely documented in previous chapters, the management of technology-based assets has been traditionally identified as the *raison d'être* of headquarters. As these assets often constitute the core of MNCs' ownership advantages, the MNC center used to be fully responsible for their generation, augmentation and protection. MNCs' massive research laboratories in the home-country have for decades defined the innovation paths of the firm as a whole, and the protection of technological knowledge has been centrally managed through both formal and informal means, spanning from the design of foreign location choices to the management of intellectual property devices.

Yet, with supply-driven motives for locating R&D activities abroad, subsidiaries tasked with innovation responsibilities have increasingly enhanced their ability to strategically manage their knowledge-based resources. One main reason explaining this trend has to be searched in the role of geographical proximity.

Technological knowledge features tacitness, social complexity and causal ambiguity. Hence, it cannot be effectively managed from a distance. Although headquarters certainly play a leading role in defining the most appropriate innovation strategies for the MNC as a whole, their ability to understand the dynamics that govern technological processes in the host locations is limited, and so is their efficacy in responding *timely, but with long-term orientation*, to the wide-range of opportunities and challenges emerging from the local knowledge network. This requires not only a comprehensive understanding of the subsidiary-level technological routines, but also a good knowledge of local actors and the ability to interact with them while accounting for the location-specific institutional environment. Because such conditions only accrue from direct interaction, which in turn is strongly facilitated by geographical proximity, subsidiaries are the only actors that can be expected to effectively manage the MNC's local knowledge base.

From the MNC internal perspective, the documented transition toward increasingly active subsidiary roles, typically associated with autonomous subsidiary-level undertakings relating to high value-adding activities, is only partially explained by the increasingly widespread decentralization of formal decision-making power in contemporary firms. In fact, the leveling process between headquarters and subsidiary roles is also powerfully triggered by subsidiary incentives.

While headquarters are the holder of ownership rights that can be exercised over subsidiaries, ownership rights are difficult to enforce when the underlying resources are based on intangible assets, such as technological knowledge or innovation expertise (Mudambi and Navarra, 2004). These assets cannot be easily transferred, substituted or imitated as they are embedded in the social context where they have been developed (Coff, 1999). Because in the networked MNC knowledge and technological capabilities are distributed, subsidiaries that find themselves endowed with such critical resources become powerful actors, as other units may be dependent on their skills.

Knowledge-based power is critical because it entitles to exert an influence over firm-wide strategic issues (Mudambi et al., 2014). However, as it is based on the subsidiary's substantive ability to contribute to the MNC technological progress rather than on any formal authority, it can be easily lost. It is therefore not surprising that subsidiaries are strongly motivated to continuously preserve and nurture their technological assets, given that these represent the source of their power.

Hence, along with external dynamics, a combination of MNC organizational factors, subsidiary accumulated skills and subsidiary intrinsic motivation explains the current patterns that see many foreign units actively engaging in the management of critical knowledge imperatives that used to fall within the exclusive responsibilities of headquarters.

While the notion of subsidiaries as strategizing actors is by now widely established in the IB literature, the way this notion reflects in subsidiaries' technological behavior is yet to be completely understood.

The investigation into subsidiaries' innovation strategy is not easy, as it requires several perspectives to be merged. While the IB literature has enabled us to capture the complexity of subsidiaries' incentives within the broader MNC network, innovation studies (IS) perspectives offer a framework for understanding the building blocks of subsidiary technology strategy.

Following this approach, it can be argued that two main objectives are critical for firms engaging in innovation processes, namely the creation and the protection of valuable knowledge-based assets (Teece, 1986, 2000). On one hand, the firm's technological portfolio has to be constantly sustained and renewed. In fact, the technological sources of its competitive advantage are not immutable but rather may change, and do so especially in technology-intensive industries. To survive and stay competitive, a firm's technology base needs to be updated and upgraded through continuous new knowledge creation.

On the other hand, technology assets have to be protected, so as to enable firms to appropriate the benefits arising from their innovative

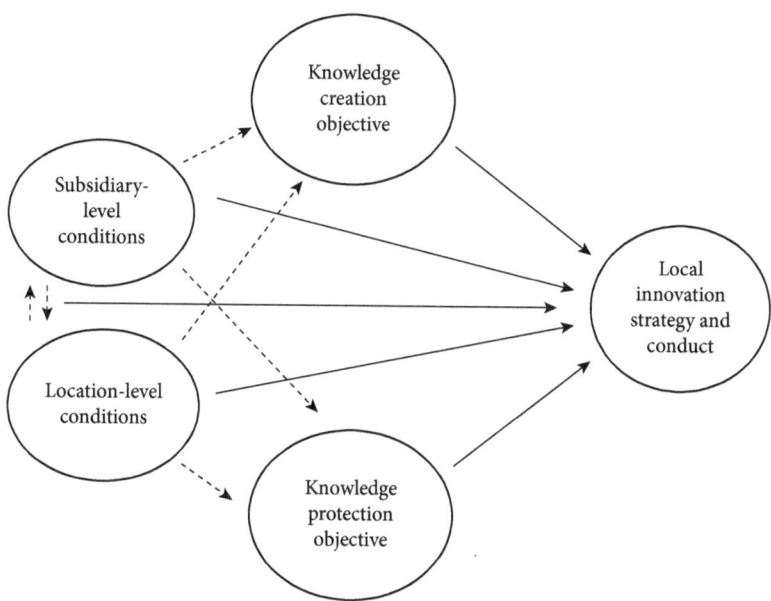

FIGURE 3.1 *The relationships between subsidiary-level and location-level conditions and objectives in subsidiary innovation strategy*

efforts. Knowledge protection is critical for the sustainability of firms' competitive advantage. In fact, as technologies get imitated, their competitive value is reduced and ultimately exhausted.

Moving these reflections to our MNC subsidiary context, we shall argue that subsidiaries that strategically manage their technological resources in their host locations are agents that "take charge" of both such innovation objectives, which we shall refer to as knowledge creation and knowledge protection (Perri and Andersson, 2014).

3.4.1 Knowledge creation

For firms' relying on technological innovation as a competitive tool, the creation of new knowledge is a persistent activity. Though valuable, established resources are destined to loose their competitive potential. Hence, firms need to continuously and resiliently renew their knowledge base.

This approach has been highlighted starting from the works on the *knowledge-creating company* and the *learning organization* (Leonard-Barton, 1995; Nonaka and Takeuchi, 1995; Simonin, 1997). Similarly, innovation literature on firms' search behavior has also been devoted to understand where firms could find new sources of ideas to expand their knowledge, come up with new exploration patterns and, in turn, develop new technologies (Chesbrough, 2003; Katila, 2002; Laursen and Salter, 2006). In other words, there is large agreement on the importance of knowledge strategic renewal for a firm's sustainable competitive advantage.

The need for knowledge creation is also critical in MNC subsidiaries. Active R&D subsidiaries are tasked with augmenting the technological home-base of the firm, and their prospects of influence-acquisition within the MNC depend on their ability to carry out such responsibility.

The process through which subsidiaries seek to advance their technology has outcomes that strongly depend on their surrounding environment and, more specifically, on their ability to gain access to the knowledge it embeds. Yet, to be redeployed to the benefit of other MNC units, the newly created knowledge has to be effectively recombined with the available knowledge inputs, and integrated in the pre-existing MNC knowledge base through appropriate interfaces that could facilitate internal circulation and reuse.

It can be argued that this process requires subsidiaries to be endowed with at least three capabilities (Phene and Almeida, 2008):

- a capability to detect relevant knowledge;
- a capability to source knowledge;
- a capability to recombine knowledge.

While the well-established notion of absorptive capacity (Cohen and Levinthal, 1990) embraces many facets of such capabilities, it is important to devote separate attention to each of them, as they are associated with critical complexities arising from the multilocation nature of subsidiaries' potential knowledge sources.

As far as the detection of knowledge is concerned, it is important to recall that subsidiaries are embedded into two networks of knowledge: the MNC internal network, which includes both the headquarters and other subsidiaries, and the subsidiary's local external network, which includes the various actors operating in the host location (Almeida and Phene, 2004).

A basic precondition to recognize knowledge resources that could be beneficial for the subsidiary's innovative activities lies in its own R&D investment (Cohen and Levinthal, 1990). Yet, despite the technical expertise subsidiaries might have accumulated, other obstacles may arise in this stage. In fact, identifying relevant knowledge in the MNC internal network requires overcoming *geographic distance* barriers. Similarly, detecting knowledge inputs in the local external environment demands subsidiaries to overcome *organizational* barriers.

In the first case, intra-organizational linkages, particularly if working on a personal level, may help subsidiaries to identify whether any of its sister units is already endowed with the knowledge it needs (Lahiri, 2010). In the same vein, the establishment of both formal and informal ties with members of the local knowledge community, such as vertical partners, competitors or other types of institutions, may help recognizing valuable technological resources in the host location.

Once knowledge has been located, subsidiaries need to deploy techniques to gain sufficient acquaintance. This step is particularly critical when external knowledge is concerned, not only because the relevant knowledge is located outside the MNC's organizational boundaries, but also because – at least in the early stage of its lifecycle – the subsidiary has an "outsider" status in the local knowledge network.

Subsidiaries may use different modes to tap into the local knowledge pools, such as hiring local engineers or entering formal collaboration agreements with local actors. Yet, overcoming the *liability of outsidership* (Johanson and Vahlne, 2009) is a necessary condition to enable an easy and convenient access to actual and prospective technology inflows.

For subsidiaries to gain an "insider" status in their host location knowledge network, thus becoming embedded, they must engage in close and recurrent interactions that instigate social networks (Rogers and Larsen, 1984) and, eventually, trust. With social networks and trust, firms learn to benefit from each other, as they conduct joint problem solving and share critical information on their own technical domains (Gulati, 1995).

Such process of gradual local integration is not automatic, though. Rather, it is likely to be activated as subsidiaries demonstrate that they possess a valuable bundle of resources, which could make them attractive partners for potential knowledge exchanges. Concurrently, subsidiaries need to develop a "reputation for cooperation". Several authors have pointed out that local actors will hardly be willing to share their technical knowledge and local insights with foreign agents if they expect to obtain nothing in return (Perri and Andersson, 2014; Schrader, 1991). Hence, subsidiaries need to exhibit a cooperative attitude toward local counterparts, using their influential resources as *bargaining chips* (von Hippel, 1987), to be able to tap local pools of knowledge.

The rationale behind such expected pattern has to be searched in the social network literature, which highlights the critical role of network structures and common third-party ties as social monitoring mechanisms (Coleman, 1988; Reagans and McEvily, 2003): because actors in a network may easily identify opportunistic actions, they can sanction such behaviors by excluding uncooperative participants from future interaction opportunities (Reagans and McEvily, 2003). Accordingly, for subsidiaries to gain an easy access to locally embedded knowledge, they must be willing to engage into reciprocal relationships with domestic actors that facilitate the acquisition of an "insider" status.

As the relevant local knowledge has finally become available to subsidiaries, it can enter the innovation process. Innovation is based on the recombination of existing pieces of knowledge with new inputs that organizations might have sourced from outside their boundaries. For subsidiaries, developing a combinative capability is not an immediate achievement, as the knowledge inputs they may need to integrate are geographically and organizationally distributed. These may originate from the headquarters, the worldwide dispersed sister units, as well as external firms in a variety of countries. Phene and Almeida (2008) suggest that the architecture of a subsidiary knowledge is fully effective when it builds upon the full-range of sources to which it may gain access. In other words, a subsidiary's combinative capability is a managerial capability, which originates through the development and continuous upgrade of complex routines through which subsidiaries combine heterogeneous technical components (Phene and Almeida, 2008) to enhance their innovation performance.

3.4.2 Knowledge protection

Along with knowledge creation, the protection of knowledge from external appropriation is the other big concern highlighted by the innovation management literature. Research has focused on the identification of techniques and instruments that may be used to isolate a firm's technological resources from the risk of imitation, such as patents and complementary assets (Shane, 2001; Teece, 1986).

Knowledge protection is also a major issue in MNCs. The relevance of this argument can be assessed in the light of established literature on entry modes, according to which the use of FDI to enter foreign markets is in fact a choice of internalization. Especially when the MNC's activities are characterized by a high degree of technological intensity, their foreign organization is more likely to be arranged through a hierarchic governance mode, which may enable not only to minimize the transaction costs related to the entry into an external market (Buckley and Casson, 1976; Hennart, 1982; Kogut and Zander, 1993), but also to keep the firm-specific knowledge within the organizational boundaries, thus limiting the risks of unwanted dissemination (Rugman, 1981).

Traditionally, the protection of the MNC's intangible assets has been considered as an exclusive responsibility of parent companies. For instance, literature has shown how headquarters design R&D location decisions cautiously to avoid excessively dangerous environments, where knowledge outflows would strongly exceed knowledge inflows (Alcácer, 2006; Alcácer and Chung, 2007; Chung and Alcácer, 2002; Sanna-Randaccio and Veugelers, 2007). Similarly, it has been suggested that the management of intellectual property tends to be centralized in the MNC's home country, thus generating *safe nests* where the most promising technological projects can be securely developed (Di Minin and Bianchi, 2011).

However, subsidiaries' transition toward an active technological behavior, and the resulting incentives to preserve the value of their knowledge in order to foster their strategic role within the MNC network, suggest that they will increasingly involve in the dynamic protection of their valuable technology.

The rationale behind this argument also lies in the idea that, just like subsidiaries find it easier to access local knowledge given their physical presence in the host location, outflows of their own technology toward local firms are equally facilitated by geographical proximity. In other words, if we accept that knowledge circulates more easily and rapidly among co-located agents, then we should also envisage that active subsidiaries would dynamically protect their technology through their own local strategies, thus accounting for potential unexpected threats emerging from the external environment.

Failing to protect the MNC technology through subsidiaries' local strategies would imply invalidating the MNCs' central protection strategy. In turn, failing to account for subsidiary-level technology protection strategies would imply neglecting a critical part of the MNC protection activities.

Hence, while proximity facilitates the creation of channels that foster local knowledge diffusion, firms are not defenseless in front of the risk of external appropriation of their technology. Although spillovers may occur unintentionally, it is not accurate to assume that MNC subsidiaries will passively accept the leakage of their knowledge to local competitors, neither it is realistic that all of them will commit the same effort in preventing this phenomenon.

Investigation into the active role subsidiaries play in the protection of their knowledge assets within foreign locations has shown that subsidiaries defend their knowledge more extensively than local firms do, by using very broad and differentiated sets of protection mechanisms, such as patents, secrecy, lead time, complex design (de Faria and Sofka, 2010) and even vertical partners' management (Perri et al., 2013).

Moreover, in support of the idea that a true strategizing activity is in place in advanced subsidiaries, empirical studies have demonstrated that these dynamically manage local protection strategies to adapt to the set of local opportunities and threats, as well as to subsidiary-internal conditions. As a case in point, in technology proficient host locations, where subsidiaries are motivated to develop linkages that allow for effective knowledge sourcing, they seek to reduce the strength of protection mechanisms so as to activate reciprocity dynamics (de Faria and Sofka, 2010).

In a similar vein, it has been showed that just like corporate managers of leading MNCs avoid co-locating in relatively backward host locations, where they would loose much more knowledge than they could gain, highly competent subsidiaries are able to make accurate evaluations of the net spillover they would experience within specific locations, thus adapting their investment into local linkages as a means to control their valuable technology (Perri et al., 2013).

The dynamic configuration of subsidiary control over its knowledge may act as the source of misalignments between foreign subsidiaries and headquarters, confirming that the social and political contexts where subsidiaries are located, along with the global environment to which they participate, often generate conflicting interests (Dörrenbächer and Geppert, 2011). Though subsidiaries may be willing to trade part of their knowledge for local resources and useful information to gain influence within the MNC, headquarters are likely to have contrasting objectives of spillover avoidance. If subsidiaries have power, then the outcome of such disputes will arise from negotiated processes, rather than being automatically imposed by headquarters.

On the whole, such reflections propose an active subsidiary role in managing local innovation that existing literature has not yet fully explored. While the dynamics of subsidiary roles and strategies have been widely investigated in general terms, IB research still has to engage in the specific account of what such evolution implies for the management of MNC technological knowledge assets.

The literature survey conducted in this chapter also points to the critical role that actors, web of relationships and wide-ranging conditions in the host location play in subsidiary innovation management and performance. Leveraging economic geography (EG) perspectives, next chapter analyzes these aspects more extensively.

4
Perspectives on Host Locations

Abstract: *This chapter employs EG interpretative lenses to analyze host locations as key contexts for MNC subsidiaries' innovation. In particular, it offers a general perspective on critical constructs such as place and space, distance and border effects and national and sub-national geographical scales. Moreover, it provides with a basic systematization of the different spatial configurations that, along the years, have been proposed for the study of innovation in the geographical space. Finally, building on recent insights calling for the integration between IB and EG in the study of MNCs in the geographical space, it argues that an approach encompassing both the IB theoretical focus on the organization and the EG emphasis on the location is especially relevant for the study of MNCs' geographically distributed innovation activities.*

Keywords: clusters; geographical space; sub-national heterogeneity; territorial systems of innovation

Perri, Alessandra. *Innovation and the Multinational Firm: Perspectives on Foreign Subsidiaries and Host Locations*. Basingstoke: Palgrave Macmillan, 2015. DOI: 10.1057/9781137555441.0011.

4.1 The relevance of locations for MNC innovation

As widely discussed in Chapters 1 and 2, traditional multinational corporation (MNC) models focus on the ownership advantages residing at the center of the firm's organization. In such frameworks, MNC competitive advantage emanates by and large from the competitive advantage of the firm's home-country (Iammarino and McCann, 2013). Along the years, these models have changed to account for the emergence of important pockets of technological capabilities in diverse areas of the globe, and for the resulting modifications in MNCs' organizational and spatial structure.

The previous chapter has highlighted issues related to subsidiary innovation strategy, pointing to the critical nexus linking subsidiaries' evolution to the host location. As subsidiaries cease to be passive implementers of headquarters' objectives to become themselves strategizing actors, their relationship with the host location becomes key for innovation processes and outcomes. Indeed, not only does the MNC's access to geographically distributed sources of competences depend on its network of foreign subsidiaries and, in turn, on their ability to tap into locally situated knowledge, but also the subsidiary's own strategic development within the firm's internal organization mainly rests on its activities and achievements within the host location.

Given the importance of host locations for both MNC-level and subsidiary-level innovation performance, the economic geography (EG) interpretative lenses become crucial to understand the interactions between organizational, technological and spatial dimensions involved in the innovation process. Accordingly, the need for integration between the international business (IB) and the EG theoretical perspectives has clearly emerged in the works by Beugelsdijk et al. (2010), Beugelsdijk and Mudambi (2013), McCann (2011) and Iammarino and McCann (2013), to mention a few, and is currently at the center of scholars' research agenda.

Approaching this issue from the IB perspective, it has been argued that two main arguments explain the importance of bridging such theories. First, the eclectic (Ownership-Location-Internalization) paradigm (Dunning, 1977), which after many decades from its original formulation is still the most widespread model for the interpretation of MNC behavior, has been updated and improved to become very effective in explaining the economic organization of geographically distributed firms, thereby allowing for an extensive understanding of both *ownership* and *internalization* advantages, but has left the *locational* dimension much less explored and defined (Beugelsdijk and Mudambi, 2013). In other words, while the investigation of ownership and internalization advantages has led to the development of very convincing answers to the questions of *why* and *how* MNCs go international, the *where* interrogative has been addressed only indirectly, and in a fragmented way (Iammarino and McCann, 2013). This is almost the contrary of what has happened in the EG field, where a sophisticated understanding of the *location* has been achieved, while *organizational issues* have been addressed with less emphasis (Cantwell, 2009).

Second, and related to the first argument, the level of geographical analysis that has been traditionally used to study MNCs' spatial behavior is often too general to disentangle the actual mechanisms linking firms to locations (Beugelsdijk et al., 2010; Beugelsdijk and Mudambi, 2013). Driven by the conventional dichotomy between home and host country, most location studies in IB literature have focused on the *national scale* to analyze the MNC geographical dimension (Iammarino and McCann, 2013). Such approach, which has been at least partially determined by data availability, limits scholars' ability to capture more fine-grained territorial phenomena that are critical to explain the interaction between MNCs and their localities, as it implicitly assumes that MNCs base their spatial choices on some "country averages".[1] While it is true that MNC location decisions involve the identification of a specific target country, MNCs are certainly not indifferent to within-country spatial heterogeneity and, in fact, they increasingly search for very specific territorial characteristics that are likely to be found only in precisely identified sub-national areas (Narula and Santangelo, 2012). Investigating MNCs' spatial behavior based on country-level, mean-based indicators does not allow to capture the complex interaction between different geographic scales and the specificities of locational features MNCs account for when taking their decisions, nor it is fully informative as we seek to understand the relationships between such firms and their localities, and the ways they evolve over time.

To address these limitations in IB research, scholars have called for a more intense leverage of EG perspectives, whose interest in the relatively under-investigated dimension of the eclectic theory, namely *the location*, may offer critical insights on how to account for spatial variation in the study of MNCs.

It has been argued that what IB scholars can learn from EG is that locations can be best analyzed by disentangling between "space" and "place" components of locations (Beugelsdijk et al., 2010; McCann, 2011). Place depicts the location-specific features of a particular geographical unit of analysis, which does not necessarily overlap with the country. Space, in very general terms, covers any source of variation that can make places heterogeneous among each other.

Although space and place are related concepts, economic geographers highlight that there are critical distinctions among them, which call for the use of tailored approaches for their analysis (McCann, 2011). For instance, one important foundation of IB literature lies in the idea that foreign business activities are more hazardous compared with domestic ones, because the *liability of foreignness* (Hymer, 1976) increases the cost of doing business abroad. Yet, while the notion of liability of foreignness is by definition associated with the existence of national borders, that is, formal discontinuities that separate a firm's business activities in space, it is often used almost as a synonymous of the concept of distance that, instead, is a continuous variable (Beugelsdijk and Mudambi, 2013).

In relation to this point, it has been recently suggested that an important distinction exists between "border effects" and "distance effects" (Beugelsdijk and Mudambi, 2013). The former take place as firms' activities cross borders where spatial transaction costs augment in a discontinuous way (Beugelsdijk et al., 2010). The latter vary in a continuous manner as distance between a firm's activities increases. While EG and regional science have mostly analyzed "distance effects", the focus of IB literature has been on "border effects", which have been largely identified with national boundaries. However, it has become increasingly clear that this is an oversimplification of reality, as a country's sub-national areas could be extremely heterogeneous among each other, too. As spatial discontinuities may and do arise even within the same country (Beugelsdijk and Mudambi, 2013), neglecting them would equal to analyze "border effects" as though they are continuous. Accordingly, to comprehensively understand the consequences of MNCs' spatial behavior, both border effects (potentially operating at different geographical scales) and distance effects need to be accounted for.

Summarizing these considerations, it can be argued that to better understand MNCs' spatial behavior, thereby gaining a more informed knowledge of locational advantages, IB studies need a higher geographical specificity. Accordingly, different geographies such as sub-national regions, cities or industrial clusters have to be considered as relevant levels of analysis, as they better inform regarding the complex set of territorial characteristics MNCs account for when calculating opportunities and risks that enter their decision-making processes. Similarly, a more detailed account of both border and distance effects is needed to understand the challenges of managing across space.

Following such interdisciplinary approach to the study of MNCs in the geographical space, the aim of this chapter is to highlight that the integration between IB and EG is particularly useful for understanding MNCs' foreign innovative activities and, more specifically, the innovative dynamics of MNC subsidiaries in their host locations.

Clearly, the most natural field of application of an integrative approach merging IB and EG is the analysis of MNC location choices (e.g., Goerzen et al., 2013). Yet, the two interrelated trends depicted in previous chapters – namely, the gradual shift of foreign locations' role from sources of market expansion or cost reduction to sources of knowledge, and the emergence of subsidiaries as active innovation developers – suggest that another primary area of investigation that requires the combination of such disciplines refers to the subsidiary local innovation process, and to the wide-ranging organizational, strategic, technological and geographical aspects it encompasses.

Indeed, while MNCs have been broadly depicted as organizations that span distances (Iammarino and McCann, 2013), capable of accessing and leveraging knowledge across geographical space, Chapter 3 has demonstrated that – given the headquarters' orchestration of the MNC's organizational units – foreign subsidiaries are the actual "spanning" entities, as theirs is the direct responsibility for setting up mechanisms for local knowledge sourcing and creation.

A more accurate representation of both subsidiaries and locations, and of the multiplicity of interactions between and around them, is critical to understand the innovation process of foreign subsidiaries, and of the MNC at large. In this regard, the fit between a subsidiary's innovation strategy and conduct and its host region locational features is important. In fact, because a two-way relationship exists between MNC subsidiaries and their host locations, their interaction pattern is the source of co-evolution processes.

This chapter offers a brief survey of the EG literature on the variety of analytical frameworks that can be used to investigate the role of local systems in innovation: it is in such systems that MNCs foreign subsidiaries interact to manage their knowledge resources.

Traditional approaches to innovation in MNCs largely consider the subsidiary as a centrally driven actor, whose behavior is imposed by the firm's head-office, and the host location as a black-box, whose features are determined exogenously. In such scenario, studying the MNCs' peripheral innovation processes is an unproductive exercise, as there are very few levers to govern. However, the evolution of IB theory on subsidiary roles and strategies on one hand, and the integration of EG perspectives on locations on the other hand, open up a fruitful avenue for additional research on active local innovation processes in MNCs' foreign subsidiaries.

4.2 Geographical systems of innovation

As suggested by Iammarino and McCann (2013), the geographical dimension of innovation systems has become central in the literature on innovation and technological change as scholars have recognized the importance of contextual factors and external knowledge sources for the processes of innovation development and diffusion at different levels. This aspect is particularly relevant as we acknowledge that there is a high degree of heterogeneity in terms of knowledge-based locational advantages across sub-national areas.

As discussed in Chapter 1, technological knowledge is inherently complex and tends to embed a relevant tacit component, both aspects that contribute to make its diffusion process geographically bounded. Moreover, innovation occurs in tight interaction with its contextual environment, as it results from the capabilities that are available in a given region that, in turn, depend upon the learning and upgrading processes developed by local innovative actors over time. In other words, technological innovation is cumulative, incremental and path-dependent (Nelson and Winter, 1982), all features that have a critical effect on its spatial dimension, thereby contributing to ascribe to geographical space a sort of selection power, which determines whether a specific location is a conducive to innovative activities (Iammarino and McCann, 2013).

Within an evolutionary economics approach, scholars have tried to understand which actors and characteristics enable locations to become a favorable context for innovation, giving rise to the fecund stream on innovation systems. More specifically, the idea that innovation is an interactive process gradually lead to the development of the concept of National Innovation Systems (Freeman, 1987; Lundvall, 1992) in which learning dynamics, depicted as institutionally embedded, constitute the basis of innovation.

The notion of National Innovation System (NIS) was intended to cover all the actors and factors that, within a country's boundaries, directly contribute to or influence the creation, commercialization and diffusion of technical change, thereby shifting the focus from the central and almost exclusive role of firms to include a differentiated set of institutions such as universities, research centers, legal and financial entities, government agencies, and so on (Lundvall, 1992). Accordingly, a critical advancement that this model offered to the study of innovation processes lies in the idea that firms' innovation processes do not occur in a vacuum, nor do they take place in isolation from other actors. Rather, innovation processes interact with a country's economic structure and institutions, such that several organizations and features relating to the research infrastructure, the education system, the legal and financial systems as well as the industrial policies may enable or constrain innovation.

Though the idea of "system" turned out to be very successful, as it pointed out that the interrelationships between elements are critical to innovation performance just like the elements themselves, and that higher-level structures influence micro-level dynamics and vice-versa (Lundvall, 2007), the focus on the national scale has been less fortunate, as its underlying assumption of within-country homogeneity soon became the object of criticism.

A first attempt to redefine the notion of NIS relates to the so-called Sectoral (SISs) and Technological (TISs) Innovation Systems, which build on the idea that while nation-states' boundaries certainly play an important role for any techno-economic field, in many cases the systemic features relating to specific sectors or technological domains are best understood if considered at geographical scales that do not coincide with the nation, but may either be local or global depending on the nature of the interactions linking actors and factors in different techno-industrial areas (Breschi and Malerba, 1997; Carlsson and Stankiewicz, 1995; Mowery and Nelson, 1999). Both SISs and TISs are at least theoretically consistent with the previously discussed idea according to which border effects may operate at different territorial levels; however, they do not take an explicit stand regarding the role of space.

The need for a more fine-grained evaluation of the systemic nature of innovation served as the trigger for the development of the *regional* approach to innovation systems, reinforced by the growing evidence demonstrating that it is at this specific geographical level that innovation is developed through social, organizational and technological interaction. Regions are critical scales where the orchestration of economic activities occurs: they operate as a meso-level between the nation-state and the individual firm (Asheim and Gertler, 2005). Accordingly, compared with TISs and SISs, the Regional Innovation System (RIS) may span more than one sector and technology; compared with NISs, it captures innovation dynamics at a sub-national level.

The RIS approach is consistent with the NIS analytical model in that it applies the NIS's fundamental elements to more circumscribed geographical scales; however, contrary to NIS approaches, it enables to account for several aspects that could not be captured at the national level, such as region-specific *patterns of industrial specialization* and *innovation performance* (Breschi, 2000; Howells, 1999), *sub-national institutions* that are critical in facilitating knowledge diffusion (Cooke and Morgan, 1994), as well as *informal* learning and innovation mechanisms that are geographically bounded (Morgan, 2004). In other words, to identify a RIS, the "top-down" elements and conditions adapted from the NIS model are not sufficient; rather, "bottom-up" features (such as dense social networks, informal rules of coordination, localized routines for knowledge sharing) need to be present to facilitate the system's functioning through an appropriate organizational and institutional set-up (Iammarino and McCann, 2013).

The lack of any of these critical features may determine the system's failure. In fact, RISs may collapse because of missing or inappropriate elements (for instance, because of the lack or the insufficient development of local universities and research institutes) and because of missing or inappropriate interactions between elements (for instance, because of the lack of integrative capabilities between universities and regional firms) (Tödtling and Trippl, 2005).

Moreover, even when the RISs exhibit the whole set of critical actors and a high degree of wide-ranging system integration, they should not be treated as self-sustaining entities (Morgan, 2004; Todtling and Trippl, 2005). In fact, the narrower the geographical boundaries of the RIS, the higher the extent of non-local linkages it has to develop to feed and enrich its internal resource base (Morgan, 2004).

4.3 Agglomeration, clusters, cities

The foregoing discussion suggests that, while the nation-state remains a highly significant geographical scale for innovation, there is a meso-level that cannot be overlooked, as it explains the widely documented within-country heterogeneity whose recognition is currently invoked by IB research.

RISs serve as important analytical frameworks to account for such heterogeneity, in that they represent the milieu where a wide range of informal institutions playing a fundamental role in knowledge diffusion/creation operate. However, RISs in turn contain smaller subcomponents that, even within restricted geographical boundaries, may exhibit characteristics that qualify them as fundamental settings of knowledge generation. In this regard, EG scholars have devoted great attention to concept of clusters, giving rise to a strand of literature that, in the last decades, has also been revitalized by more strategic perspectives (Porter, 1990, 2000).

To understand the importance of clusters, it is useful to recall the literature on spatial agglomeration, which is rooted in the work of Alfred Marshall (1890, 1920). According to the author, firms' location dynamics are influenced by the existence of intra-industry positive externalities that are local bound, thereby accruing only to firms that agglomerate in a circumscribed local area. Such agglomeration economies arise from three main phenomena:

1. The existence of *knowledge spillovers*, occurring as tacit knowledge circulates, both intentionally and unintentionally, among co-located industry members;
2. The availability of *specialized inputs*, such as services, intermediate inputs or infrastructures, which become accessible at large scales given the size of the local market;
3. The presence of *specialized skilled labor*, which tends to accumulate abnormally in a given location as industry-specific local competences develop over time.

In all three cases, geographical proximity and specialization enable more efficient search and transfer processes. No matter if the objective of the search is the solution to a technical problem, a particular components or a new worker, firms that co-locate in an industrial cluster are likely to find it more easily, compared with non co-located organizations.

Marshall's praise of specialization finds support in subsequent works by Arrow (1962) and Romer (1986) that, integrated in the so-called Marshall-Arrow-Romer (MAR) model (Glaeser et al., 1992), overall suggest that knowledge is sector-specific, and that spillovers primarily involve firms operating in the same or similar industry.

The antagonist perspective to Marshall's view of clusters is traditionally identified in the work of Jane Jacobs (1961) proposing that, along with co-location, it is the diversity of industrial activities, rather than its specialization, to enhance learning processes and act as the real trigger of local innovation. According to the author, the simultaneous presence of a variety of industrial activities in the same geographical area is conducive to new ideas that can be used in different contexts. Hence, co-located firms benefit from the exchange of knowledge arising from diverse technical bases, and the most critical sources of spillovers are external to a firm's industry. In terms of spatial scale, the work of Jacobs (1961) refers to large metropolitan areas; accordingly, she identifies cities as the major sources of innovation, where diversified production activities, availability of wide-ranging services and infrastructures and proximity to markets generate what have been labeled "urbanization economies" (Hoover, 1937).

While empirical research has remained inconclusive on the dispute between Marshallian, intra-industry externalities and Jacobian, inter-industry economies, recent work has suggested that if a synthesis has to be found between the two perspectives, it is useful to rely on the notion of "related variety" (Boschma and Iammarino, 2009; Iammarino and McCann, 2013). This concept is important to understand the effect of a region's production structure on innovation. In fact, it helps to specify the nature of "regional diversity", which can be either related or unrelated. Arguably, knowledge spillovers will only occur when there is a sufficient degree of relatedness, or cognitive proximity, among the diverse technical bases embedded in a region, as this allows for the emergence of synergistic exchanges of knowledge and competences. It should be noted here that the notion of *related variety* is not completely extraneous to the work of Jacobs (1961) either. In fact, her thinking acknowledges that, while diverse, local industrial activities need to be complementary to enable effective cross-fertilization. Yet, just like high cognitive distance hinders knowledge sharing because of communication barriers, cognitive convergence reduces the potential for local learning and interaction, eventually leading to lock-in scenarios (Boschma and Iammarino, 2009).

This suggests that an unambiguous relationship between the degree of industrial diversity and innovation performance at the local level cannot be easily identified, given the complex facets underlying the notion of "regional diversity" (Iammarino and McCann, 2013). However, while both specialization and unrelated diversity encompass risks, local specialization in related variety seems to be the right compromise for learning and innovation (Boschma and Iammarino, 2009).

When considering the sources of local creativity, it is useful to recall that clusters are not closed entities, because extra-regional relationships can bring additional variety within their boundaries. These linkages are key to complement the cluster's local *buzz* thereby avoiding technological lock-in (Bathelt et al., 2004). The full-range of potential linkages that connect one location with all other locations in the world has been termed *connectivity* (Lorenzen and Mudambi, 2013). Connectivity enables infusions of external knowledge that provide clusters with the novelty and variety that are required to nourish local innovation processes, particularly if the knowledge sources reside in a variety of foreign countries. This channel for extra-local knowledge sourcing is particularly important for emerging-market locations, which are often characterized by a poor and low-quality technological base that needs to be nurtured with more advanced knowledge inputs.

As far as the mechanisms for connectivity are concerned, it has been argued that innovative agents in a specific location can link to peers in other locations mainly through "pipelines" and "personal relationships". While the latter foster connectivity through individuals leveraging mutual social proximity, such as family-based relations or friendships, pipelines are organization-based, deliberately established channels that aim at maximizing the transfer of resources across geographically dispersed locations. In mature clusters, the organizations engendering and maintaining these pipelines are typically domestic firms, often MNCs (Gertler and Levitte, 2005; Trippl et al., 2009). However, MNCs act as key pipeline enablers also in emerging-country clusters through their local subsidiaries (Patibandla and Petersen, 2002). This is consistent with the idea that a cluster's leading firms may serve as gatekeepers, bridging the local context to the outside world and channeling extra-local knowledge inputs from various sources within the system (Giuliani and Bell, 2005).

Despite the critical importance of connectivity for clusters' innovative performance, its potential beneficial effects do not display automatically, but rather depend on the cluster's technological profile. Indeed, just like to the benefits of internal variety require relatedness to occur, clusters are able to make a productive use of external knowledge only if provided with sufficient absorptive capacity (Asheim and Isaksen, 2002) and with a matching knowledge base (Boschma and Iammarino, 2009). In other words, related variety – or cognitive proximity – serves as an enabler of synergies for both local-local and local-extra-local combinations of knowledge source-recipient.

The idea of relatedness also emerges in the view of Porter (1990, 2000), who considers clusters as geographical areas comprising firms, and other associated institutions, operating in a range of interconnected industries, and whose competitiveness relies on four major driving forces: input conditions, demand conditions, related and supporting industries, and the contextual factor governing firms' strategy and rivalry.

The distinctive trait of his approach lies in the positive role ascribed to local competition, which separates it from the MAR model where local rivalry hinders innovation by engendering too high risks of knowledge leakage. In Porter's view, cluster firms both compete and cooperate, and the degree of sophistication of their competitive approaches affects the cluster's vitality.

In turn, these firms' innovation processes are nourished by means of three fundamental mechanisms. First, clusters include specialized entities that are dedicated to the creation and circulation of information, firms endowed with buyer knowledge as well as forward-looking customers, which enable cluster firms to detect changes in buyers' needs more rapidly and effectively compared with firms that are located in isolation. Second, given the flow of technical knowledge and the concentration of specialized providers of services, components and skills, cluster firms are in a privileged position not only to uncover, but also to implement new technological opportunities. Third, because of geographical proximity, cluster firms may observe and monitor competitors' strategic moves and accomplishments, thereby being stimulated to identify original approaches to distinguish themselves from rivals.

Overall, Porter's perspective emphasizes the role of rivalry as the driving force to constant innovation and cluster renewal. Dynamic competitive pressure acts as a powerful engine that – combined with geographical proximity – exposes cluster firms to competitors' successes pushing them to reinforce their own advantage. Conversely, the diffusion of homogenous competitive approaches within clusters is seen as dangerous, as it hinders the emergence of new ideas leading to inertial developmental paths and rigidities. Despite this lock-in risk, Porter depicts clusters as the ideal territorial configuration for innovation and learning processes. In this framework, the cluster's geographical scale depends on the distance at which different types of interactions among the cluster members occur. Therefore, a cluster could potentially overlap with a city, a region or even span the national boundaries.

Though the scale of analysis remains an open question, to which admittedly it would be difficult, or even wrong, to give a conclusive answer, there is growing recognition that while clusters should be characterized primarily in local terms, their interaction with higher levels of spatial analysis cannot be overlooked as these critically contribute to the cluster's evolution, both constraining or nurturing their success (Wolfe and Gertler, 2004). In other words, local clusters need to be put in context, and considered as nested within increasingly broader innovation systems, both at the regional and at the national scale, and connected to global knowledge networks. This approach is fully consistent with the previously described model of border and distance effects associated with geographical space/place.

Recent urban growth trends highlight the increasing importance of the urban scale (Iammarino and McCann, 2013). By facilitating wide-ranging transactions through the availability of advanced infrastructure and diverse resources, large cities offer fundamental opportunities for increasing the overall efficiency of local economic activities. People and organizations are motivated to agglomerate together in urban areas by the prospects to benefit from extensive human capital interaction and continuous new knowledge generation (Berry and Glaeser, 2005). Yet, as the foregoing discussion on connectivity suggests, what determines a city's economic and innovative productivity is not (only) its size (Iammarino and McCann, 2013); extra-local connections, including the city's ability to enter global linkages, is also a critical driver of performance (Amin and Cohendet, 2004; Hannigan et al., 2015). Along with a high degree of internal cohesion, successful cities also exhibit an inherent capability to interact with similar locations spanning national borders, thereby participating to global networks.

Such capability, that enables cities to become key knowledge centers, is closely related to their accessibility and permeability, but also depends on the nature of co-located agents. Not surprisingly, MNCs' location behavior plays a fundamental role in explaining why some cities become "global" (Iammarino and McCann, 2013). Given their network of subsidiaries located in different foreign countries, MNCs are prime developers of international linkages, thereby serving as progenitors of local and global knowledge flows that, together with a cosmopolitan atmosphere and a munificence of advanced producer services, contribute to convert cities into hubs of worldwide webs of linkages (Goerzen et al., 2013).

The role MNCs play in cluster development emerges particularly in the context of non-traditional locations, namely peripheral and emerging-country locations. For instance, first-mover MNCs locating in the Catania province of Sicily have activated mechanisms of strategic isomorphism and local generic resource agglomeration, thus generating path-dependent dynamics of FDI (foreign direct investment) inflows and transforming shallow resources into an emerging cluster (Mudambi and Santangelo, 2014). Similarly, the formal relationships linking MNC subsidiaries to headquarters triggered the creation of the Bangalore IT cluster through the installment of linkages with local firms and the development of human capital and technological capabilities (Patibandla and Petersen, 2002).

Peripheral areas and emerging-country locations share some characteristics (Santangelo, 2009). For instance, their industrial and economic context is relatively immature, their innovation processes are unsophisticated and their degree of integration with the world economy is limited, partly due to their less privileged geographical position and scarce endowment with transportation and communication infrastructures (Santangelo, 2009). Yet, emerging-country locations are more precarious contexts for innovation, as the wider institutional framework in which they are embedded is usually more underdeveloped, thereby challenging the interactive nature of innovation processes (Scalera et al., 2015).

Despite these weak locational features, MNCs involvement in these areas is increasing at a very fast pace. This suggests that MNCs are striving to seek for sources of competitive advantage in non-core locations as a way to face increased global rivalry. However, it also shows that such locations are indeed endowed with pockets of resources to which – quite obviously – MNCs ascribe great value. In the case of emerging-country locations, these resources are typically identified in the substantial pool of skilled labor and expertise available at a relatively low cost (Lewin et al., 2009) that combine with highly growing but untapped local markets. On the other hand, peripheral locations' attractiveness usually lies in that the limited availability of sophisticated resources is coupled with government subsidies that increase the overall efficiency of local operations (Mudambi and Santangelo, 2014).

The increasing attention MNCs are devoting to emerging countries in particular, which more and more become the target of their knowledge-intensive FDI, contributes to assign them a privileged position in the current geography of innovation, and requires scholars to take a closer look at the characteristics of innovation in such contexts.

It has to be recognized that while advanced MNCs increasingly establish R&D facilities in these countries, either to satisfy the local demand or to employ local resources to satisfy global markets, local firms tend to be still involved in processes of technological catch-up. Indeed, faced with fierce competition from advanced foreign rivals, emerging market firms are striving to improve their capabilities and ultimately reach the global technology frontier.

This process is eased by the opportunities offered by the disaggregation of value chains, which allows for the gradual involvement of previously isolated locations (Mudambi, 2008). Global value chains have not only encouraged worldwide linkages, but they have also altered the resulting networks' configuration. Traditionally, global innovation networks were concentrated in advanced areas of the globe. However, the increasing participation of non-core locations is downplaying the dominance of advanced-world economies. Technological catch-up tends to follow a staged process in which emerging country firms first engage in technical collaborations with MNCs, then integrate into the industry value chain and finally become able to generate new knowledge through internal R&D (Kumaraswamy et al., 2012). The co-existence of highly heterogeneous actors pursuing diverse and often contrasting objectives makes the study of innovation in these countries extremely interesting but, at the same time, challenging. In fact, the lack of comprehensive and reliable firm-level data, together with the novelty of the phenomenon, heighten the barriers to the development of an acceptable understanding of the processes of knowledge sourcing, creation and protection in which both local and foreign firms are involved.

Regardless of the degree of development of the host location, it can be argued that when establishing abroad, MNCs adjust to both formal and informal local institutions and simultaneously contribute to the institutional development of the host environment (Cantwell et al., 2009). Similarly, they access to local knowledge sources and also contribute to the knowledge base of host territories, thereby sustaining local learning in a mutually reinforcing firm-location innovation process. Hence, MNCs and locations cannot be analyzed as though they are static. Firm strategies, institutional change and the growth of local capabilities are interdependent and contribute to each other's long-term development (Meyer, 2004). Awareness of such co-evolution processes allows to gain a dynamic view of the relationship between MNCs and host locations, as well as to reconcile the location-centric imperative of EG perspectives and the organization-centric focus of IB theory.

4.4 Place and space: locational features for MNC innovation

The arguments reviewed in this and in previous chapters have suggested that, because not all locations are the same when it comes to the systemic interactions they may enable and to the learning opportunities they embed, geography becomes a fundamental source of competitive advantage. Contrary to single-location firms, MNCs can benefit from geographical variety through their unique ability to span distance by setting up subsidiaries in different foreign areas. The identification of this critical MNC capability has fostered a prolific stream of studies on foreign knowledge sourcing and creation (e.g., Almeida and Phene, 2004; Frost, 2001; Singh, 2008), in which EG perspectives on locations in general, and on clusters in particular, have been used and re-interpreted (Porter, 1990).

However, there seems to be a widespread agreement now that early views of clusters have been over-optimistic, and their developmental potential somewhat exaggerated (Martin and Sunley, 2003). Similarly, the use of cluster literature in IB research has been slightly naïve, in that it assumed that MNC presence in foreign clusters could only lead to knowledge-based benefits. In other words, clusters were considered as relatively passive environments to which MNCs could automatically access to source valuable knowledge through the establishment of their subsidiaries. It soon became clear that this view was inadequate, as clusters are instead active and dynamic milieu, where knowledge flows bi-directionally. Following this logic, co-location, and the resulting agglomeration externalities, are not necessarily positive for all firms, but rather depend on firm-specific traits such as capabilities.

The oligopolistic deterrence paradigm developed by Shaver and Flyer (2000) explains that co-located firms simultaneously gain and contribute knowledge to the cluster in which they are established. Technologically leading firms tend to avoid locating in clusters because, by doing so, they would loose more than they could gain, thus reducing the competitive gap between them and less-developed rivals.

Extending this logic, it can be assumed that location choices across the value chain are governed not only by agglomeration benefits but also by competition costs (Alcácer, 2006), arising when firms compete in the same geographic market, and materializing through strategic behaviors such as entry deterrence or exit threats. In such framework, R&D features strong co-location dynamics: compared with other firm activities, it is more likely to benefit from agglomeration effects (especially owing to spillovers and specialized labor) and less likely to be affected by competition costs. Yet, as agglomeration enables both knowledge inflows (positive spillovers) and outflows (negative spillovers), more capable firms are motivated to isolate, rather than to agglomerate with rivals, in order to reduce knowledge leakage and preserve their competitive advantage.

These results remain consistent even when accounting for different sources of local knowledge. Indeed, while leading MNCs tend to locate close to academic and governmental knowledge sources that minimize knowledge outflows and offer opportunities for the access to sophisticated technology, laggards are mostly attracted by locations with a high level of industrial innovative activities, where they can source more "accessible" knowledge (Alcácer and Chung, 2007). Similarly, while MNCs avoid co-locating with domestic companies, unless these possess some comparative advantage, they tend to agglomerate with other foreign firms, which are perceived as being more technologically advanced (Mariotti et al., 2010).

According to this approach, if we abstract from the structure and the nature of the relationships among the cluster firms, we should expect industrial clusters involving dominant players either failing to develop or growing as "concentrations of mediocrity" (McCann and Mudambi, 2005). Indeed, technologically capable firms would shy away from localities where they are destined to obtain a negative net knowledge flow. While this perspective offers a theoretical justification for the empirical evidence showing that many of the largest companies prefer to locate their high value-added activities in isolation rather than to agglomerate (Cantwell and Santangelo, 1999), it does not explain why we observe many successful clusters, where even leading MNCs establish their competence-creating subsidiaries, and where interactive learning and innovation foster locational competitiveness.

EG perspectives have tried to reconcile this puzzle by analyzing the nature of knowledge and inter-firm relationships within clusters. Using a transaction-costs perspective, it has been suggested that spatial agglomerations are not homogeneous in terms of governance modes and participating firms' characteristics, but rather can be differentiated among three typologies, namely the "pure agglomeration", the "industrial complex" and the "social network" (Gordon and McCann, 2000). The pure agglomeration features the presence of atomistic agents in no way linked by loyal relationships, while the social network equals trust-based relationships à la Granovetter (1973) to intra-organizational hierarchies thus assuming the absence of opportunism. In the middle of these somewhat extreme models, the industrial complex encompasses agents that make an investment to gain membership to the system with the aim of developing stable linkages. In contrast to the pure agglomeration and the social network, both inconsistent with the presence of large firms capable of leveraging their organization to internalize knowledge, the industrial complex model seems to fit well with the idea of oligopolistic MNCs participating to the cluster with the objective to establish long-term relationships involving intentional knowledge sharing with similar or smaller firms (Iammarino and McCann, 2013). While recognizing the existence of negative spillovers, this view confirms that industrial clusters may develop as environments where firms, including MNCs, are willing to locate and eventually share their own technology-based assets, motivated by the prospect to obtain local knowledge in return.

Recognition of firms' heterogeneity is also critical to explain the dynamics of knowledge exchange within clusters. Giuliani (2007) argues that knowledge is not symmetrically distributed within such territorial units, but rather diffuses depending on the firms' distinct knowledge bases and the resulting relational dimensions, such that more capable firms – which local counterparts recognize as fundamental knowledge reservoirs – are likely to enjoy more privileged positions in the local knowledge network.

This perspective has also been brought within the IB literature with the physical attraction logic (Cantwell and Mudambi, 2011) that, in contrast to the strategic deterrence view's emphasis on proximity-driven knowledge spillover risks, primarily highlights the role of local knowledge sourcing opportunities. According to this viewpoint, leading MNCs do agglomerate into areas of intense industrial concentration because their dominant technological position expedites the acquisition of insider statuses, thereby ensuring a facilitated access to knowledge inflows. Conversely, technology laggards are deterred, as they lack the necessary resources to develop productive ties with sources of local knowledge.

Summarizing these observations, what emerges is that when assessing their location choices, MNCs are aware of the complex set of knowledge-based implications potentially arising from geographical proximity, which include both knowledge sourcing opportunities and spillovers threats. This leads to differences in firms' propensity to agglomerate. In turn, cluster internal features play a role in this process. There are indeed several factors that contribute to explain which type of knowledge will circulate in specific territorial units, which types of firms will find it useful to agglomerate in these areas and which arrangements and governance modes are required to gain access to locally embedded knowledge. Moreover, locations are not static, but rather evolve depending on the nature and behavior of co-located agents and extra-local linkages.

This approach, finally encompassing both IB theoretical focus on the organization and EG emphasis of locations within a general IS lens, may offer steps forward in uncovering the dynamic interaction involving different spatial industrial configurations and MNCs, thereby providing more comprehensive understandings of MNC innovation.

Note

1. As an example, geographic distance is often measured as the distance between capital cities, while economic distance as difference between countries' GDP per capita.

5
Integrating Perspectives

Abstract: *This chapter integrates IB, IS and EG perspectives to propose a comprehensive framework for the study of geographical dispersal of R&D activities within MNCs. Such analytical framework suggests accounting not only for the micro-level of MNC headquarters and the macro-level of home/host countries, but also for the more fine-grained microlevel of MNC subsidiaries and the meso-level of sub-national host locations, such as clusters. This approach has the potential to enrich the investigation into geographically dispersed R&D activities through an accurate account of actors and conditions that are more directly involved into decentralized processes of innovation. The chapter concludes by applying the proposed framework to the topic of FDI knowledge spillovers to host locations.*

Keywords: analytical framework; FDI spillovers; integrated theoretical approach; multilevel analysis

Perri, Alessandra. *Innovation and the Multinational Firm: Perspectives on Foreign Subsidiaries and Host Locations.* Basingstoke: Palgrave Macmillan, 2015. DOI: 10.1057/9781137555441.0012.

5.1 Integrating international business, innovation and economic geography perspectives

The literature survey conducted in Chapter 2 has highlighted the critical importance of multinational corporation (MNC) headquarters in the decision-making process leading to research and development (R&D) internationalization, and in the orchestration of foreign R&D units. Yet, Chapter 3 has contended that beyond the substantial influence of headquarters' R&D strategies, subsidiaries play a critical role in the management of knowledge-based assets. In particular, applying innovation studies (IS) logics to MNC foreign subsidiaries, it has been suggested that subsidiaries are subject to two different stimuli: a knowledge creation imperative, arising from the willingness to take advantage from the opportunities of learning, knowledge sourcing and capability development embedded in the host location, and a knowledge protection imperative, originating from the need to safeguard their competitive technology from the risk of local leakage (Perri and Andersson, 2014).

As both the ability to source local knowledge and the ability to protect technological assets depend on the degree of interaction with co-located agents, a tension can be expected to emerge between these knowledge imperatives (Perri and Andersson, 2014). In other words, it can be argued that – all else equal – more embedded subsidiaries will be able to source a higher and more sophisticated amount of local technology, but will also need to reciprocate such inflows with some of their own knowledge. Clearly, both internal and external influencing factors can enter the stage and change the balance between knowledge creation and knowledge protection imperatives. In turn, the way in which tension linking these imperatives evolves is likely to affect the subsidiary's choices, local strategies of technology management and conduct.

The focus on subsidiary innovation strategy and local behavior is critical to understand MNC innovation performance. In fact, beyond locating at the "right place" (Iwasa and Odagiri, 2004) and deploying effective orchestration capabilities, which clearly fall within MNC headquarters' responsibilities, the outcome of foreign R&D also critically depends on how locational advantages are managed in stages that are subsequent to the location choice, a task that is mandated to subsidiaries.

Shifting the focus from the MNC organization to the host location, Chapter 4 has leveraged recent insights emanating from both international business (IB) and economic geography (EG) perspectives to argue that, when analyzing MNC local innovation processes and outcomes, it is pivotal to account for the bulk of actors, capabilities and conditions that characterize spatial industrial configurations at different scales. While IB has long focused on the country as the relevant spatial level of analysis to study the MNC geographical behavior, thus neglecting sub-national heterogeneity and the complex set of distance and border effects operating at different geographical scales, this chapter has highlighted that, even within the same country, locations are heterogeneous along many different dimensions that matter for MNC innovation, including the nature and degree of technological sophistication of agglomerated agents and the transactional logics governing their interaction. These dimensions strongly affect the dynamics of knowledge circulation and evolution within spatial entities. Far from implying that higher-level geographical scales are irrelevant, the approach proposed in Chapter 4 suggests that sub-national territorial entities have to be analyzed in consideration of the internal locational features that more directly influence the local dynamics of knowledge diffusion and creation that matter for MNC innovation, but also as nested within broader innovation systems and connected to global knowledge networks.

Locations compete with each other to attract high value-added foreign direct investment (FDI) and gain the status of higher-order territorial entities. This process may encompass the development of "knowledge generating inputs" (Audretsch and Feldman, 1996) and of favorable socio-institutional conditions, potentially transforming clusters into centers of locational excellence, such that a ranking of the attractiveness of geographical locations emerges. In turn, MNC headquarters evaluate the extent to which locational features match their knowledge objectives and the organizational and internalization capabilities they may leverage to orchestrate local innovation processes, and act accordingly. Therefore, locational features of different spatial configurations and MNCs' corporate level strategy and behavior are two essential building blocks for the analysis of MNC innovation. Only their integrated account enables to understand the MNC-location mutual reinforcing patterns of interaction and, ultimately, the influence of such interactions on MNC innovation.

Though this approach is useful in that it encourages the integration of the micro-level of the firm and the meso-level of territorial innovation systems or clusters, it needs to be pushed forward to ensure that all relevant actors and conditions that explain MNC innovation are accounted for. More specifically, it is pivotal to combine the insights developed in

Chapter 3 regarding the role of foreign subsidiaries with those discussed in Chapter 4 regarding the heterogeneous conditions and dynamics of host locations.

MNC subsidiaries play a critical role in the firm innovation management processes. Indeed, by leveraging geographical proximity, they are the only MNC nodes to be able to implement informed, responsive and timely actions in the host location. Subsidiaries actively and dynamically create new knowledge by exploiting locational factors and protect their own technological assets within the host location, thereby managing a complex trade-off between these imperatives.

Therefore, headquarters' careful design of location choices is not the only channel through which MNCs are able to govern local bi-directional knowledge flows. Rather, MNC capability development comes to depend on diverse elements, including not only the headquarters-level strategy and structure, but also the subsidiaries' strategy and behavior within the systems of innovation in which they are established. MNC subsidiaries' local behavior, in turn, is not without consequences for host locations, as suggested by co-evolution literature (Cantwell et al., 2009).

Summarizing these insights, we shall contend that R&D internationalization can be comprehensively understood only by integrating the whole set of levels of analysis that influence its occurrence and evolution, as well as the ways in which these levels interact with each other to ultimately determine the outcomes of innovation processes. Previous literature has mainly focused on micro- and macro-levels, captured respectively through the account of MNC headquarters' and home/host countries' roles. Integrating IB and EG perspectives within a general IS angle, this book calls for a more inclusive analytical framework covering also the more fine-grained micro-level of MNC subsidiaries and the meso-level of sub-national locations, thereby allowing for a fuller understanding of the complex interplay between local and global dimensions.

Too often we still refer to general concepts of *MNCs* and *countries* when analyzing how innovation is managed in presence of geographical dispersal. This approach overlooks the fact that MNCs are complex organizations composed of both headquarters and subsidiaries, and that countries are broad systems whose internal space is characterized by high heterogeneity. Our understanding of geographically dispersed innovation processes in MNCs can greatly benefit from a more precise identification of the relevant actors, geographical scales and conditions that directly or indirectly enter innovation processes.

Therefore, beyond the critical role of MNC headquarters and home/host country influences, investigation into geographically dispersed R&D activities need to be enriched with an accurate account of those actors and conditions that are more directly involved into decentralized processes of innovation, namely the subsidiary and the sub-national host location. It is their mutual interaction that spurs outcomes whose impact exhibits on both subsidiaries and host locations, up to the headquarters and the MNC at large.

As previous chapters suggest, each level of analysis is characterized by a combination of diverse dimensions (firm, location, technology,

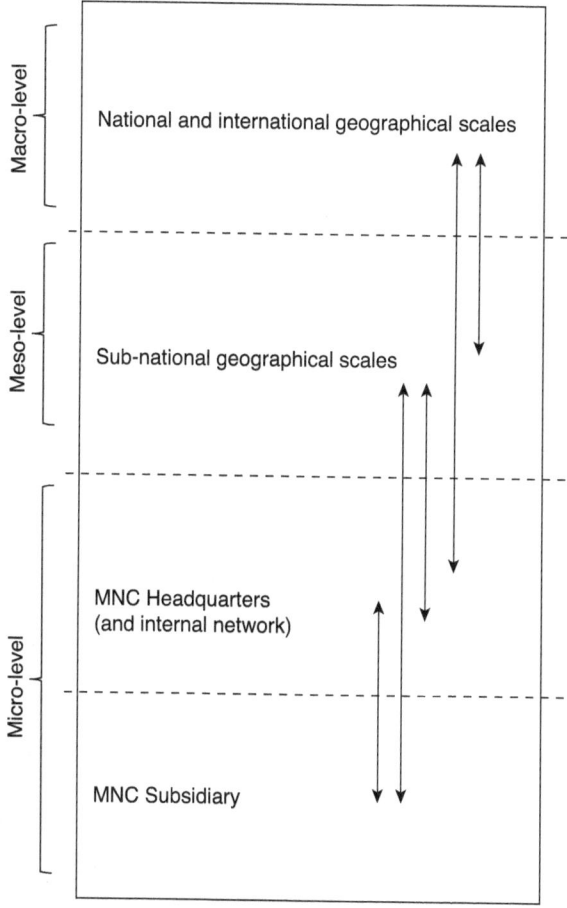

FIGURE 5.1 *A multilevel approach to the analysis of geographically distributed innovation in MNCs*

individual, etc.) that, in turn, encompass a high degree of heterogeneity that contributes to make each decentralized innovation process complex and multifaceted. Though difficult to model and measure, such multi-level heterogeneity is the basis for understanding the differentiated outcomes of R&D internationalization projects.

In the following exposition, the proposed analytical framework will be employed for the analysis of a key issue in the wide literature on R&D internationalization, namely FDI spillover to host locations. When MNC headquarters establish their subsidiaries abroad, host locations experience a broad set of implications. A nurtured stream of literature has analyzed the broad topic of FDI externalities. Given the focus of this book, we shall focus specifically on knowledge spillovers, occurring when subsidiaries' technological knowledge flows outside the firm boundaries, either intentionally or unintentionally, to be internalized by local firms.

We contend that the analysis of this specific consequence of R&D internationalization provides an appropriate application field for our framework for two critical reasons. First, because most existing studies on this topic have adopted a country-level approach, with the aim of investigating the conditions under which FDI can contribute to host country development. Second, because research on the micro-level determinants of FDI knowledge spillovers has primarily analyzed the role of headquarters-level decisions, thereby leaving subsidiary-level strategic perspectives largely unexplored.

The approach developed in this book suggests that actors and conditions characterizing more proximate territorial entities than the host country are relevant to understand the dynamics of knowledge diffusion, and that – given their incentive set – subsidiaries are not indifferent to the risk of local knowledge leakage (Perri et al., 2013). Hence, according to our framework, both sub-national host locations and subsidiary-level strategy and conduct should more directly enter FDI spillover research.

5.2   The case of FDI spillovers to host locations

When MNCs establish their subsidiaries abroad, a wide array of effects involves both the MNC host and home-country. These effects have long attracted the attention not only of scholars, but also of governments and policy makers, interested in the potential developmental impact of FDI on host countries. Literature has identified *product market effects*, occurring when the decision to set up a FDI leads the MNC to change the quantity of goods it buys or sells in its home and host country, and *factor market effects*, arising when the MNC modifies its demand for capital and labor in its host economy and at home (Barba Navaretti and Venables, 2004). Product market effects cover changes in the intensity of competition and market supply in the host economy, and in output levels at home. For example, foreign investment aimed at replacing imports by local production to better supply the local market could increase competition in the host country, thus fostering consumer welfare through price reduction, but at the same time could reduce output at domestic plants. On the other hand, factor market effects include changes in capital flows, in the demand for labor and in labor skill composition, both in the host and home-economy. For example, foreign investment could raise the demand for labor in the host country, but simultaneously increase the demand for skills at home, with consequences on wage levels in both cases (Barba Navaretti and Venables, 2004).

Beyond *product-market* and *factor-market effects*, a third class of FDI effects can be identified in FDI spillovers. Before defining FDI knowledge spillovers, it is useful to stress that the constructs of spillovers and externalities do not exactly overlaps. Externalities take place when FDI generates outcomes that become available to other actors at no cost. Spillovers are externalities that occur between specific foreign and local agents, thereby presupposing the existence of some sort of formal or informal relationships between them (Morrissey, 2012).

FDI-related externalities can be classified as (1) pecuniary/non-pecuniary or (2) inter-industry/intra-industry (Eden, 2009). FDI *pecuniary externalities* derive from buyer-supplier relationships to which the MNC participates and influence *the supply or demand conditions* for local consumers or other firms (Dunning and Lundan, 2008). They usually take place through vertical (backward and forward) linkages, that is, business relationships with local suppliers and distributors.

Non-pecuniary externalities occur when the activities of MNC subsidiaries in the host location influence the local firms' technological endowment. They may be either horizontal or vertical. Indeed, even if vertical linkages essentially generate pecuniary spillovers, they may also foster processes of adaptive learning (Jindra et al., 2009).

Inter-industry effects concern firms operating in industries other than that in which the MNC operates. Typically, they occur through supply-chain relationships (Rodriguez-Clare, 1996) and may encompass both pecuniary and non-pecuniary effects. In the latter case however, the technological knowledge shared within vertical relationships with local partners may also benefit the MNC's rivals (Spencer, 2008) through indirect contacts mediated by common suppliers or distributors.

Intra-industry effects involve those firms belonging to the same industry as the MNC. Different mechanisms facilitate their occurrence. For instance, the "demonstration effect" appears when local firms learn how to replicate MNCs' production, marketing and/or organizational practices, as a consequence of their exposure to MNC activities (Blomström and Kokko, 1998). Competitive information about the MNC's technology may also flow toward local firms through the channel of inter-firm labor mobility (Glass and Saggi, 2002). Another intra-industry mechanism is the "competition effect", occurring when FDI intensifies the competitive pressure on local companies, thereby encouraging them to improve their product and process technologies as a way to survive to foreign rivalry (Blomström and Kokko, 1998).

Except for the latter mechanism, which induces local firms to make a more productive use of *their own knowledge* to respond to foreign competition, FDI non-pecuniary externalities, also known as *technology- or knowledge-spillovers*, materialize in the (either intentional or unintentional) *flow* of knowledge-based resources *from one organization to another* (Perri and Peruffo, 2015). As the actors involved in such knowledge flow, that is, the foreign subsidiary and the local firms, are often linked by competitive relationships, the management of spillovers is a vital activity for subsidiaries, as loosing control over valuable knowledge may deteriorate their relative competitive standing. Symmetrically, from the perspective of local firms, FDI knowledge spillovers may serve as means to upgrade their technological assets thus enhancing their ability to compete with foreign investors (Perri and Peruffo, 2015).

Therefore, from both the MNC and the host location perspectives, understanding the drivers of FDI knowledge spillovers is pivotal. Literature has widely documented variables and conditions that contribute to explain this phenomenon. Yet, as the foregoing discussion suggests, the shortcomings that have been highlighted in previous chapters regarding levels of analysis and theoretical lens that are still missing in the study of phenomena related to R&D internationalization also characterize the literature on FDI knowledge spillovers and make it an ideal topic to analyze in the light of the multilevel framework proposed in this book.

Conceptualizing the antecedents of FDI spillovers through the integration of IB, IS and EG perspectives requires reviewing the most relevant theoretical and empirical findings on the subject, and systematizing them in light of the insights regarding the active role of subsidiaries and host locations, and in consideration of different relevant levels of analysis, namely the macro-level of the country, the meso-level of sub-national host locations and the micro-level of headquarters and subsidiaries. This enables to understand what we know and what we should know about this phenomenon.

5.3 Antecedents of FDI knowledge spillovers

The main theoretical reference for the analysis of FDI spillovers' antecedents is the eclectic theory, which suggests that MNCs are typically endowed with ownership advantages. Ownership advantages provide MNCs with sophisticated knowledge and capabilities that can be deployed abroad through the transfer to foreign subsidiaries. When this technology infusion from the MNC "core" to decentralized subsidiaries takes place, the interaction between foreign firms and local companies may give rise to spillovers (Haskel et al., 2007).

Macro-level research on the antecedents of FDI spillovers has been very fecund. Conversely, meso-level and more micro-founded determinants of spillovers have been investigated less extensively. In the following section, FDI spillovers' antecedents at different levels are discussed to uncover the most critical research gaps.

5.3.1 Macro-level perspectives

Macro-level research on the antecedents FDI spillovers has resulted in a wide range of studies, most of which investigating home- and host country features (Ford et al., 2008), and specifically the effect of the *technological gap* which separates them. Theoretical models predict that in the presence of a higher technological distance between the home- and host country, the potential for learning via FDI will be greater (Findlay, 1978). However, empirical evidence on this postulate is mixed (Haddad and Harrison, 1993; Kokko, 1994; Jordaan, 2005; Li and Liu, 2005; Takii, 2005).

The lack of unambiguous findings can be at least partially explained by accounting for the role of absorptive capacity (Glass and Saggi, 1998; Perez, 1997; Wang and Blomström, 1992). As verified by empirical research (Ben Hamida and Gugler, 2009; Blalock and Gertler, 2009; Blalock and Simon, 2009; Borensztein et al., 1998; Damijan et al., 2013; Girma, 2005; Jordaan, 2005; Kemeny, 2010; Liu et al., 2000, 2009; Li and Liu, 2005; Liu and Buck, 2007), while large technological gaps are potentially associated with relevant learning opportunities, neither do these materialize automatically, nor do they accrue to all local firms homogeneously (Görg and Strobl, 2001).

To benefit from FDI knowledge spillovers, local firms must possess a sufficient level of technological capabilities (Blalock and Simon, 2009; Girma, 2005; Girma et al., 2008; Huang et al., 2012; Liu et al., 2000; Li and Liu, 2005) that may enable them to detect valuable external knowledge, internalize it and employ for commercial objectives (Cohen and Levinthal, 1990). In other words, absorptive capacity helps local firms to *realize* potential FDI spillovers (Meyer and Sinani, 2009).

A different viewpoint on the role played by the technological gap proposes that it influences the type of knowledge MNCs will choose to transfer to host countries (Glass and Saggi, 1998). In presence of large gaps, headquarters will likely convey the less advanced technology to foreign affiliates, thereby limiting the potential for spillovers.

It has also been suggested that, in presence of excessively low or high levels of absorptive capacity, local firms will hardly benefit from spillovers, either because they are unable to internalize foreign knowledge or because they already possess state-of-the-art technologies (Girma, 2005; Huang et al., 2012). More generally, it can be argued that moderate levels of absorptive capacity allows for the greatest local learning potential (Meyer and Sinani, 2009), as these ensure that the knowledge sources and recipients are technologically proximate enough to fulfill effective knowledge exchanges.

FDI knowledge spillovers also vary across industries, depending on their technological intensity (Alvarez and Molero, 2005; Buckley et al., 2007) and on host country competition. The latter in particular has been ascribed a dual role. On one hand, high competitive pressure may encourage MNCs to endow subsidiaries with superior technology thereby generating greater learning opportunities (Wang and Blomström, 1992). On the other hand, it may induce MNCs to protect their asset more extensively, thus raising barriers to local knowledge diffusion (Fosfuri et al., 2001).

Macro-level cultural and institutional factors are also expected to influence the relevance of FDI knowledge spillovers. For instance, cultural, social and legal differences between the home- and host economy modify foreign firms' incentives to develop linkages with local partners (Rodriguez-Clare, 1996). Culture also influences the individual's knowledge-sharing approaches (Michailova and Hutchings, 2006). Knowledge that is not coherent with the recipient country's culture will encounter societal, institutional and legal barriers in the transfer from headquarters to subsidiaries (Hennart and Larimo, 1998), thus reducing the spillover potential.

The influence of host financial markets and commercial regulations has also been investigated. Well-developed financial systems help host countries to take advantage from FDI (Alfaro et al., 2004, 2009, 2010; Choong, 2012). On the other hand, when investing in host countries characterized by restrictive trade regimes, MNCs limit their local operations to low value-added tasks, thus generating lower levels of local linkages (Belderbos et al., 2001; Kohpaiboon, 2006).

5.3.2 Meso-level perspectives

Compared with the large attention devoted to macro-level antecedents of FDI knowledge spillovers, literature on meso-level drivers is much more limited. Most related research has dealt with the debate on whether FDI acts as a trigger for cluster development and, in turn, for spillover effects or, conversely, clusters develop endogenously, mainly as a result of local agents' collective core competences, thereby becoming attractive also for knowledge-seeking FDI (De Propris and Driffield, 2006).

Early attempts to account for the role of sub-national heterogeneity have investigated the differential ability to reap FDI knowledge spillovers of assisted vs. non-assisted regions (Driffield, 2004). Moreover, research has explored how the dominant type of innovative activity of local economic areas in the US influences MNC location choices as a way to maximize net spillovers (Alcácer and Chung, 2007).

To explore how different forms of agglomerations heterogeneously combine with the presence of MNCs, meso-level research has demonstrated that the highest potential for FDI spillovers is associated with the presence of marshallian industrial districts (Menghinello et al., 2010).

More recently, the effect of the structure of competition as a determinant of heterogeneous MNC governance choices of innovative activities and, in turn, as a factor influencing FDI knowledge spillovers has been analyzed at the level of the cluster (Alcácer and Zhao, 2012), suggesting that leading MNCs that co-locate with direct market rivals foster more extensive internal linkages across locations to reduce spillover risks.

Despite these selected contributions, research should more widely account for other sources of meso-level heterogeneity that could significantly affect FDI spillover patterns, including the host location spatial structure, its geographical position, the degree of openness and global connectivity, and the social norms governing the subsidiary's local relational network. Critically, research on the sub-national institutional drivers of FDI knowledge spillovers has not comprehensively disentangled the potentially heterogeneous influence of formal and informal institutions on spillovers, and the way these interact with other spillover drivers.

A sparkling example of the contribution meso-level perspectives may offer to the study of FDI knowledge spillovers is offered by the debate on absorptive capacity. Despite the wide recognition of the critical role it plays in the dynamics of local spillovers, this construct has conventionally been depicted at the country-level and has not been conceptualized in ways that account for factors and relations that facilitate knowledge exchange at the local-level.

Although absorptive capacity critically depends on the learning investment of individual firms, it is likely to be powerfully influenced also by the conditions that characterize the immediate environment where such firms operate (Narula and Driffield, 2012). In most countries, these conditions strongly differ across regions. Region-specific institutional factors, such as the education system, the cultural atmosphere and the infrastructures for codifying and disseminating technological knowledge (Meyer, 2004) contribute to firms' accumulation of learning capabilities in fundamental ways. For instance, weak learning institutions in the host location may hamper local firms' knowledge acquisition, even when the knowledge transfer from MNCs occurs intentionally. Failing to account for these features is dangerous as it could lead to misinterpret empirical results.

Similarly, the structural properties, governance modes and social norms of the local business network to which firms belong deeply affect their ability to undertake the wide range of activities required to effectively absorb knowledge (Eapen, 2012). Yet, we still need to fully understand how these and other relevant meso-level factors may contribute to explain the patterns of FDI spillovers to host locations.

5.3.3 Micro-level perspectives

Moving the analysis to the micro-level allows to account for the fact that FDI knowledge spillovers result from the interactions among single firms and, eventually, among individuals within these firms. Starting from the headquarters role, literature has mainly related spillovers to the location choices of technology-intensive FDI, as tools deployed by MNCs to isolate their technology from the risk of local dissemination.

As recalled in previous chapters, MNCs often choose to agglomerate with foreign peers to gain access to local or sector-specific knowledge (Mariotti and Piscitello, 1995). However, despite facilitating learning from local firms, co-location also spawns outward spillovers. Accordingly, empirical research on FDI knowledge spillovers confirms that learning processes and demonstration effects work better among co-located agents (Barrios et al., 2006; Driffield, 2006; Thompson, 2002), although a few exceptions exist (Potter et al., 2002).

MNCs select their location seeking to maximize inward knowledge flows while minimizing outward spillovers, thereby accounting for their stronger or weaker competitive position relative to local firms' technological endowment (Alcácer and Chung, 2007), as well as for the type of potential co-located agents (Alcácer and Zhao, 2012) and for heterogeneous local knowledge sources (Chung and Alcácer, 2002).

Beyond location choices, other headquarters-level decisions, such as those relating to entry modes (Belderbos et al., 2001) and degrees of foreign ownership (Ethier and Markusen, 1996) have been investigated as potential spillover antecedents. Greenfield FDI have been found to be associated with greater local learning potential because, compared to acquisitions, they are more likely to embody the headquarters' sophisticated technology (Branstetter, 2006; Liu and Zhou, 2008).

In principle, organizing FDI through a wholly-owned subsidiaries should help to exclude local firms from the access to MNC technology. However, the empirical evidence is mixed, suggesting either that full ownership is actually more effective in isolating the MNC knowledge compared to joint venture arrangements (Tian, 2010) and to joint domestic-foreign ownership, which encourages more extensive local linkages (Javorcik, 2004; Javorcik and Spatareanu, 2008), or that local participation to subsidiaries' ownership does not in reality affect spillovers (Blomström and Sjöholm, 1999).

The influence of the nationality of ownership of foreign investors has also been variously documented (Javorcik and Spatareanu, 2011; Zhang et al., 2010). In particular, it has been showed that different types of ownership advantages, resulting from the MNC's nationality of origin, spawn heterogeneous spillover effects (Buckley et al., 2002, 2007; Wei and Liu, 2006).

Finally, research has emphasized the importance of the motivations behind the establishment of foreign subsidiaries (Morrissey, 2012). FDI motivations vary, often in accordance with MNC strategies, and hence may impact the likelihood of knowledge diffusion within the host location. Empirical research has demonstrated that technology-exploiting FDI, featuring established technology-based advantages, enhance domestic productivity by bringing new technology to the host country, while technology-sourcing projects do not generate spillover because, unlike introducing technology, they seek to gain access to local knowledge and expertise (Chung, 2001; Girma, 2005; Driffield and Love, 2007). Moreover, while both horizontal (market-seeking) and vertical (efficiency-seeking) FDI encompass positive spillovers, the effects of horizontal subsidiaries are more pronounced, as they spawn greater local embeddedness (Beugelsdijk et al., 2008).

Though being innovative in their theoretical contributions, as they more thoroughly incorporate strategic considerations in the study of FDI knowledge spillovers, these works feature a major weakness in that most of them infer the motivation for FDI (a firm-level construct) from the analysis of industry-level variables, thereby distancing this literature stream from the actual level of analysis it refers to.

Moreover, as the MNC structure and strategies become increasingly complex and sophisticated, FDI spillover research needs to recognize that novel FDI motivations may arise, that motivations evolve to reflect the changing role that particular host locations play for MNCs (Narula and Driffield, 2012), and that also subsidiaries may have their own motivations, driving their local behavior and modifying the overall expected FDI impact, in terms of knowledge spillovers.

Conventional works on FDI knowledge spillovers did not conceive subsidiary-level heterogeneity, as they consider FDI as *homogeneous* and *exogenous* elements of the MNC expansion strategies (Driffield and Love, 2007). In fact, traditional "pipeline" models of FDI spillover suggest that foreign subsidiaries serve as passive conduits for transferring knowledge from headquarters to host locations; in other words, they are depicted as leaky repositories of the MNC centrally developed knowledge (Marin and Bell, 2006). However, these models fail to consider that, while the establishment of subsidiaries abroad distributes the MNC knowledge across geographical space, it does not necessarily lead to any leakage of knowledge beyond the firm's organizational boundaries (Blomström and Kokko, 1998). To understand when such knowledge leakage is more likely to happen, it is essential to accept that, beyond the transfer of knowledge from headquarters, subsidiaries may heterogeneously protect knowledge from external appropriation risks, as well as cultivate their own capability set, often in evolution with host locations (Narula and Driffield, 2012).

Embracing this viewpoint, empirical research has started to explore the role of *subsidiary characteristics and technological behavior*, validating the prediction that subsidiary-level heterogeneity matters for the intensity of FDI knowledge spillovers.

For instance, it has been found that subsidiaries that acquire technology on international markets are also more likely to convey knowledge to the host location (Veugelers and Cassiman, 2004). Literature has also showed that only subsidiaries performing technologically creative undertakings positively contribute to the advancement of the host economy (Marin and Sasidharan, 2010), and that their contribution evolves with the time spent in the host location (Giroud, 2007).

The influence of subsidiary role, autonomy and mandate has also been documented (Giroud et al., 2012; Jindra et al., 2009; Spencer, 2008). This stream shares some similarities with older regional economics perspectives seeking to analyze how MNC headquarters' assignment of roles and responsibilities to local affiliates could affect host locations' patterns of growth (Hood and Young, 1976, 1988; Young et al., 1994). For instance, early research suggests that while local firms benefit more from affiliates tasked with a wider set of value-added functional responsibilities (Young et al., 1988), externally controlled plants, with their low degree of local integration, exert narrow or even negative developmental effects on host countries (Watts, 1981). Shifting to IB-driven approaches, more recent studies confirm that subsidiaries characterized by creative roles and greater autonomy intensify the relationships with local actors and, particularly, the recourse to local sourcing, thereby activating important opportunities for spillovers (Giroud et al., 2012; Jindra et al., 2009; Spencer, 2008).

As the foregoing literature review suggests, the emergence and increasingly widespread diffusion of subsidiary-level research has triggered a number of insightful views on the effect of subsidiary characteristics on FDI knowledge spillovers. However, the role that subsidiary strategy, intended in terms of *the management of its knowledge assets*, may play in the spillover process has long remained unexplored (Perri and Andersson, 2014).

This has happened despite IB scholars' call for a greater account of MNC strategy at the subsidiary-level (Cantwell and Mudambi, 2005) and of the emergence of polycentric perspectives depicting MNCs as political coalitions where power is diffused among different MNC units (Ferner et al., 2005), and where subsidiaries autonomously engage in complex relationships with other internal nodes and local actors to pursue their own strategic objectives (Kristensen and Zeitlin, 2005).

Following these views, and in accordance with the proposed analytical framework, we shall suggest that subsidiaries can *strategically manage* their knowledge in the host location, thereby generating heterogeneous patterns of spillovers.

Recent empirical research has started to offer some insights on this topic, for instance by analyzing subsidiary knowledge protection strategies (de Faria and Sofka, 2010; Tian, 2010). Because subsidiaries recognize the knowledge appropriation risks potentially arising from the interaction with local firms, they actively select a combination of formal and informal protection tools (de Faria and Sofka, 2010).

However, as suggested in Chapter 3, local companies do not only represent a threat for subsidiaries, but they also embed opportunities in terms of knowledge sourcing (Cantwell, 1989). Accordingly, the analytical framework proposed in this book suggests that managing knowledge in the host location entails protecting subsidiary knowledge, while concurrently exploiting opportunities for local sourcing.

Protection-oriented approaches will make subsidiaries more closed toward the external context, thus narrowing spillover channels and simultaneously limiting local interaction opportunities. Conversely, explorative approaches will make subsidiaries open and permeable to the host location's knowledge dynamics, thus ensuring access to local technology while simultaneously increasing the opportunities for spillovers (Perri and Andersson, 2014).

The balance between these knowledge creation and knowledge protection imperatives is not easy to reach, as it is likely to be influenced by varying internal and external conditions, depending on which subsidiaries may favor one objective over the other (Perri and Peruffo, 2015). As a case in point, dense network structures generate powerful social monitoring mechanisms, which in turn make subsidiary cooperative behavior more likely (Eapen, 2012).

In this regard, recent research has argued that subsidiaries actively seek to manage this complex trade-off in consideration of wide-ranging influencing factors, for instance by deploying narrower protection tools when seeking to profit from knowledge embedded in the host location (de Faria and Sofka, 2010), and by adjusting their investment into local linkages to both competitive conditions in the host environment and the value of subsidiary capabilities (Perri et al., 2013).

These results suggest that, just like firm-level heterogeneity influences headquarters' location choices to limit spillovers, subsidiary-level heterogeneity drives more competent subsidiaries to exert a higher degree of control over their knowledge-base, given the relatively higher competitive loss they would experience in case of spillover (Perri et al. 2013; Perri and Andersson, 2014). Similarly, just like headquarters are able to evaluate opportunities and risks of different host countries' locational features, subsidiaries may be sophisticated enough to perceive the implications of varying conditions in the host locations, and to adapt their local behavior accordingly.

Failing to account for this critical subsidiary skill hinders our ability to evaluate the actual antecedents and moderators of FDI spillovers, with obvious consequences for MNCs', local firms' and policy makers' strategies.

5.4 Implications for the study of FDI spillovers to host locations

The brief literature review conducted in the light of the multilevel analytical framework proposed in this book confirms that the most widely debated drivers of FDI knowledge spillovers are those operating at the macro-level of the country and at the micro-level of MNC headquarters. Therefore, it suggests that more attention needs to be paid to the meso-level and subsidiary-level variables and mechanisms governing the FDI spillover process. Indeed, the reasoning developed in previous chapters suggests that, although the account of headquarters-level

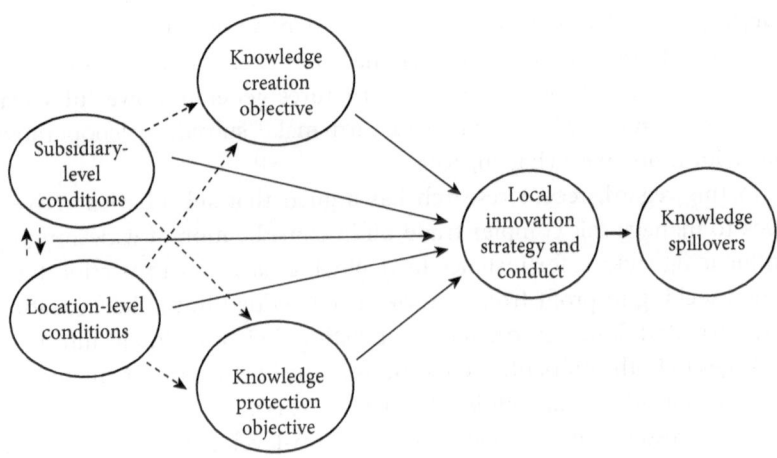

FIGURE 5.2 *Subsidiary strategy and conduct and local knowledge spillovers*

decision making is critical in predicting FDI spillover potential, the size of "*realized*" spillovers will likely depend on actors and conditions that are closer to where spillovers actually occur.

A more comprehensive analysis of the subsidiary-level conditions under which these actors voluntarily activate channels of local spillovers, or raise barriers to the dissemination of their technological resources would complement existing literature.

Similarly, the role of sub-national features such as the concentration and technological profile of co-located actors, the presence of technological infrastructures, and the integration into global innovation networks, could be explored to offer more insights into meso-level drivers of FDI knowledge spillovers. More detailed meso-level analyses could also help to better understand the spatial dimension of the spillover phenomenon, thus reconciling existing mixed results proposed by the literature stream on the *geographical scope* of FDI knowledge spillovers, that seeks to understand the extent to which knowledge outflows from FDI are spatially constrained (Driffield et al., 2004; Driffield, 2006; Chang and Xu, 2008; Crespo et al., 2009; Hong and Sun, 2011; Ben Hamida, 2013; Potter et al., 2002).

Moreover, despite the nascent literature stream finally merging subsidiary-level variables with conditions in the external environment, a careful account of how micro and meso-level features combine to influence local spillover patterns is still missing. This is a relevant area of research because, as our analytical framework suggests, the subsidiary technological profile and strategic approach cannot be expected to work in isolation from meso-level features in determining the overall FDI impact in terms of knowledge spillovers. Despite the relevance of these aspects, little is known regarding the meso-level conditions under which subsidiaries will embrace more protective or more open approaches to the management of their knowledge assets, and regarding how these interact with other subsidiary-level variables.

On the whole, the analytical framework proposed in this book appears to be consistent with the multilevel architecture of the FDI knowledge spillover phenomenon and capable of offering a more holistic view of this complex subject (Perri and Peruffo, 2015). While the most critical research gaps involve meso-level and subsidiary-level perspectives, our analytical framework also suggests that future research should not omit to study this phenomenon as nested in established country-level and headquarters-level contexts.

The multilevel approach to FDI knowledge spillovers also calls for the adoption of different relevant theories, which are needed to account for both contextual and strategic variables, as well as for the interaction among them. Such approach implies significant changes also for empirical research on the topic, which to date has either worked with a specific level of analysis, largely overlooking other levels, or covered multiple methodological levels in the theoretical development, but empirically tested only the headquarters or country level (Perri and Peruffo, 2015).

6
Concluding Remarks and New Research Directions

Abstract: *This chapter summarizes the most relevant arguments developed in this book. Starting from the acknowledgment of the persistent importance of geographical proximity, particularly for innovation-related matters, it explains why R&D internationalization and, in turn, the geography of international innovation, is a relevant research topic despite the seemingly limited magnitude of this phenomenon. It emphasizes the importance of conceptualizing subsidiaries as strategizing actors in the management of local technological resources and host locations as contexts that actively affect the dynamics of MNC decentralized innovation processes. Finally, it reviews the features of the proposed multilevel analytical framework reconciling IB, IS and EG perspectives, and discusses the potential avenues for new theoretical and empirical research.*

Keywords: blended approach; empirical design; research opportunities; theoretical development

Perri, Alessandra. *Innovation and the Multinational Firm: Perspectives on Foreign Subsidiaries and Host Locations.* Basingstoke: Palgrave Macmillan, 2015. DOI: 10.1057/9781137555441.0013.

Concluding Remarks and New Research Directions   125

6.1 Concluding observations

This book systematizes and critically discusses the research on the geographical dispersal of innovative activities in multinational corporations (MNCs). Investigation into this issue proceeds in a context marked by two influencing debates regarding the role of geography and the relevance of the research and development (R&D) internationalization phenomenon.

Taking stock of insights arising from different and often contrasting perspectives on the importance of geographical proximity and co-location for the distribution and evolution of economic activities at large, the book suggests that while geographical proximity cannot be considered as a self-sustaining catalyst for technological upgrade, it serves as a key enabling factor for face-to-face interaction and, in turn, for effective formal and informal knowledge-based transactions that foster the emergence of new ideas and solutions, thereby triggering innovation.

Clearly, not all knowledge is the same. As opposed to explicit knowledge, tacit knowledge has a personal character, and it is embedded in a specific context such that, when applied to a different setting, it often loses its strategic value. Transferring tacit knowledge requires a process of direct interaction between individuals. Moreover, some knowledge components are cultivated and held collectively through experience and spread through specific socially-embedded channels. Increasing levels of tacitness, specificity and complexity exacerbate the barriers to knowledge absorption. Under this condition, geographical proximity may prove to be insufficient to ensure knowledge transfer, and increasingly broader sets of proximity (social, cognitive, etc.) become necessary. Although firms can improve their technological base simply by capturing bits of explicit knowledge that were previously unknown and that do not need particular dimensions of proximity to be transferred, access to more tacit knowledge tends to have greater developmental potential. In other words, it could be argued that there are levels of knowledge, levels of proximity and levels of technological potential. The critical role of geographical proximity in this realm lies in that it is highly conducive to *other* dimensions of proximity, such that it may directly and indirectly enable the flow of resources that have very high potential for technological upgrade. Clearly, this does not mean that orchestration of tacit knowledge across distance is impossible, as some firms have already

DOI: 10.1057/9781137555441.0013

demonstrated to have the advanced capabilities to do so (Cantwell and Santangelo, 1999).

Moving to the research on the relevance of R&D internationalization, the book highlights that the patterns of this phenomenon may strongly vary across countries and sectors. R&D has long lagged behind the much more globalized architecture of other business functions, such as production and purchase activities. However, reading this evidence as the validation of the hypothesis according to which R&D is a case of non-globalization, as some authors have pointed out (Patel and Pavitt, 1991), could be an oversimplification. Compared with other activities, innovation features an array of specific characteristics that make it a very complex and challenging field of management. There are reasons to believe that moving R&D abroad is not as "easy" as moving production abroad, even in presence of very powerful locational pulls. In this regard, this book embraces widespread views according to which the relevance of R&D internationalization needs to be evaluated in qualitative, rather than quantitative, terms (Cantwell, 1995b). Innovation becomes authentically international when, in addition to standardized and adaptive activities, also more creative and technology-driven functions start to be conducted in foreign locations. In other words, while a significant portion of foreign R&D activities is still motivated by the need to adjust products and processes to the specificities of international markets, the evidence showing that a significant number of foreign subsidiaries do perform *creative* technological activities in their host locations provides a valid argument in support of the thesis claiming that R&D internationalization is in fact a relevant trend, which has a bearing on the most strategic areas of the firm development. Although adaptive R&D is likely to be more geographically widespread and quantitatively significant, because it follows the needs of production and commercialization functions and, in turn, of foreign markets, more sophisticated forms of R&D can be expected to involve a selected number of geographical areas that embed distinct combinations of locational advantages. In fact, while markets are (almost) everywhere, world-class technological resources are by definition rare. Accordingly, while market-driven innovative activities are likely to be more pervasive, both in geographical and in quantitative terms, technology-driven foreign investment tend to be much more selective. However, this should not advise toward the interpretation of creative, technology-driven foreign investment as a trivial phenomenon, because it is its inherent nature, as well as its underlying motivation, that drive such pattern, thus limiting its magnitude. On the contrary, the very existence of technology-driven foreign activities, which documents MNCs' quest for geographically dispersed technological resources, is a sufficient reason to investigate the wide-ranging facets of this phenomenon.

Granted that the geographical dispersal of MNC innovative activities is a relevant area of investigation, this book explains that, despite the rich literature on the topic, there are still theoretical perspectives and levels of analysis that need to be more deeply scrutinized and effectively integrated to deliver a realistic and comprehensive account of the processes leading to specific innovative outcomes.

This book proposes a novel approach, in accordance with the recent call for the integration of international business (IB) and economic geography (EG) theoretical lenses in the study of MNCs activities in the geographical space (Beugelsdijk et al., 2010; Beugelsdijk and Mudambi, 2013; Iammarino and McCann, 2013). It shows that such integration is particularly relevant when innovation-related issues are to be analyzed, given the critical role that geography exerts on processes of learning, knowledge creation and knowledge diffusion, and the relevance of geographically dispersed pockets of advanced knowledge for MNCs' technological development. Moreover, to ensure a comprehensive understanding of innovation in MNCs, it proposes to blend IB and EG perspectives with innovation studies (IS) insights, which contribute to better account for the specificities of knowledge and technology-based dynamics.

IB literature on the geographical distribution of innovative activities in MNCs has long focused on the role of headquarters, an approach resulting from established models of MNCs depicting the parent company as the sole engine of technological development and holder of strategy-making responsibilities. The same literature has offered a largely superficial conceptualization of locations, which have been theoretically identified and empirically operationalized with the national scale, due to conventional views of MNCs' ownership advantages as arising from the firm home-country.

Interestingly enough, in the meantime, EG literature developed complementary weaknesses and strengths. In fact, while leaving essentially unexplored the logics and dynamics behind the economic organization of geographically distributed firms (Cantwell, 2009), it developed a deep understanding of locations and, in turn, of the geographical concepts of place and space (Beugelsdijk and Mudambi, 2013).

Reconciling these synergetic theoretical perspectives while leveraging necessary IS insights is critical to understand the management of innovation across geographical space that, as suggested by this book, is an inherently multilevel phenomenon, which calls for an appropriate integration of necessary levels of theory and levels of analysis. More precisely, this approach suggests that the macro-level of the country, the meso-level of sub-national locations and the micro-level of both MNC headquarters and subsidiaries, as well as the interaction among these levels, need to be accounted for when seeking to understand the drivers, influencing factors and implications of the geography of international innovation.

The book also suggests that the levels where there is more need for additional theoretical and empirical development are the micro-level and the meso-level. At the micro-level, the analysis of the subsidiary technological profile, strategy and conduct is critical to fully specify the firm-specific antecedents and consequences of MNCs' innovation processes. Literature has by now widely acknowledged that subsidiaries are not the passive implementers of headquarters-level strategy depicted in traditional models of MNCs. Yet, this view of subsidiaries may still offer wide-ranging opportunities of application in the study of innovation-related matters. In fact, both theoretical reasoning and anecdotal evidence suggest that subsidiaries are increasingly active in these fields, as they recognize that the management of their technological resources will strongly influence their strategic evolution and even survival within the MNC network. Moreover, the intense leverage and deployment of locational knowledge-based advantages may enable them to gain power in the face of higher-level lines of management. In other words, subsidiaries act as truly *strategizing actors* in the management of innovation and technological resources: they are provided with their own incentives and are aware of the internal and external opportunities and challenges they need to manage to pursue their own objectives. Failing to account for subsidiary technological strategy and behavior entails dangerous misinterpretation risks, including the risk of ascribing to headquarters merits or demerits that in reality fall within the subsidiary responsibility, or the risk of overlooking the interactions that specific firm-internal or external factors may create with subsidiary-level conditions or decisions.

At the meso-level, the proposed blended approach enables to introduce a more nuanced view of locations that is essential to ensure a better understanding of MNCs geographical innovative behavior. First, it is important to account for the fact that both border effects and distance effects influence the MNC's organization of its geographically distributed innovation activities. Second, it is essential to consider that even within the same country, locations may be highly heterogeneous along many different dimensions that matter for MNC innovation, including the nature and degree of technological sophistication of agglomerated agents, the transactional logics governing their interaction, the presence of knowledge generating factors, and the socio-institutional infrastructure, to name a few. These factors strongly affect the dynamics of knowledge circulation and evolution within spatial entities. Failing to account for them exposes the risk of overlooking some of the potentially most critical drivers of MNC managerial choices at different levels, and some of the dynamics that more directly determine the outcome of MNC innovation processes.

In other words, both the subsidiary and the sub-national host location are the source of heterogeneity that, if neglected, may lead to wrong decisions and interpretations. Moreover, just like subsidiaries, locations are not passive. They grow in combination with co-located actors, such that processes of subsidiary-location co-evolution can be detected, where subsidiary-level and location-level variables and conditions interact to ultimately determine direct and indirect implications of innovation processes.

On the whole, such reflections suggest that the key players of MNC's geographically distributed innovative activities, such as searching, transferring, sourcing, negotiating, protecting and creating knowledge, are subsidiaries and host locations. Therefore, to fully account for their critical role, it is vital to shift the focus of analysis at the levels where they operate and can be effectively observed, that is, the levels where decentralized innovation actually happens.

This does not imply that higher-level geographical scales (such as the country) or managerial roles (such as the headquarters) are irrelevant, as both sub-national host locations and subsidiaries need to be studied in consideration of the wider contexts in which they are nested. The approach developed in this book suggests that, while these are important building blocks that certainly influence the outcomes of innovation processes, such processes are likely to be more directly affected by sub-national territorial entities and subsidiary strategy and conduct. More in general, it can be argued that only the integrated account of the multi-level set of actors and conditions involved in geographically dispersed innovation activities enables to understand the MNC-location mutual reinforcing patterns of interaction and, ultimately, the effect of such interactions on MNC innovation.

6.2 Potential research opportunities

The proposed multilevel approach is likely to foster the development of more holistic conceptualizations of phenomena relating to the geographical distribution of innovative activities, in which different theories belonging to heterogeneous disciplines including IB, EG and IS harmonically integrate.

Such approach is likely to spawn fundamental changes also for the empirical treatise of this realm. To date, most empirical literature has either stressed a specific level of analysis, without accounting for other levels, or incorporated multiple methodological levels in the conceptual development, but empirically tested only the country-level or the headquarters-level. Moreover, constructs used to represent relevant features of specific levels of theory have often been operationalized through variables computed at different levels of analysis, hardly able to capture the specificities of the original constructs. A multilevel theoretical perspective on the geographical distribution of MNC innovative activities instead demands a consistent empirical approach, whereby the scope of empirical investigation expands to fit with broader theoretical frameworks (Hitt et al., 2007), and levels of theory carefully align with levels of analysis and measurements (Rousseau, 1985).

Therefore, from both a theoretical and an empirical perspective, the proposed analytical framework encompasses opportunities for new research and for a better understanding of phenomena related to the management of geographically dispersed innovative activities in MNCs. From a theoretical standpoint, a more pervasive leverage of EG perspectives on host locations, along with a refined account of IB and IS perspectives on subsidiary technology strategy, may uncover new insights on the antecedents, influencing factors, dynamics and implications of decentralized innovation processes occurring in MNCs' host locations, at different levels of analysis. This offers opportunities to develop more inclusive paradigms for understanding the complexity of organizing innovation across geographical space.

From an empirical viewpoint, a more accurate design of testing strategies, capable of accounting for the multilevel architecture of the phenomenon, may help to better model underlying theoretical conceptualizations thereby disentangling the actual role of relevant constructs, which in the past have suffered from level-related confusion, and simultaneously allow for the emergence of previously overlooked interactions across levels.

More specifically, some of the research questions that could emerge from the application of this approach are the following: How do border and distance effects influence subsidiary knowledge sourcing and knowledge protection strategies? How do meso-level knowledge-based infrastructures influence subsidiary innovation performance? What are the most critical sub-national sources of knowledge for subsidiary innovation? How and to what extent do sub-national socio-institutional factors moderate the relationship between subsidiary innovation's inputs and outcomes? Does co-evolution between subsidiaries and host location always generate positive consequences for subsidiary innovation? How does the subsidiary degree of centrality within a cluster change with changing levels of cluster global connectivity? How do headquarters react to increasingly autonomous subsidiary strategy-making in technological issues? Which informal knowledge protection tools do subsidiaries leverage to safeguard their knowledge in specific host locations? Which locational features influence subsidiaries' leverage of knowledge protection tools? Which governance modes and social norms in the host location mostly stimulate subsidiary patterns of local interaction? How do subsidiaries find a balance between contrasting (positive and negative) characteristics of higher-level and lower-level spatial configurations? Under which dimensions do subsidiaries and host locations need to fit with each other to activate beneficial patterns of co-evolution? Can subsidiaries actively and consciously select among heterogeneous types of local partners to maximize net spillovers?

Scholars should particularly explore issues relating to the geography of international innovation in non-traditional contexts, such as emerging-country locations. These areas are in a continuous evolution and the speed at which institutional, economic and social conditions change combined with the increasing involvement of both domestic and foreign MNCs in the local innovative dynamics, offers an ideal setting for investigation.

Obviously, IB scholars in particular should be aware that advancing our knowledge of the geography of MNC innovation by expanding levels of theory and levels of analysis is not an easy task. It requires adopting a multilevel, nested approach to the study of relationships between subsidiary and co-located agents, understanding the structure and properties of host locations, using a wide set of appropriate research methods and gaining access to wide-ranging data, whose limited availability over the last decades has strongly constrained our ability to achieve a systematic understanding of the phenomenon. Hopefully, this book offers a useful starting point to address such complex but highly promising areas of research.

Bibliography

Agrawal, A., Kapur, D., and McHale, J. (2008). How do spatial and social proximity influence knowledge flows? Evidence from patent data. *Journal of Urban Economics*, 64(2), 258–269.

Alcácer, J. (2006). Location choices across the value chain: How activity and capability influence collocation. *Management Science*, 52(10), 1457–1471.

Alcácer, J., and Chung, W. (2007). Location strategies and knowledge spillovers. *Management Science*, 53(5), 760–776.

Alcácer, J., and Zhao, M. (2012). Local R&D strategies and multilocation firms: The role of internal linkages. *Management Science*, 58(4), 734–753.

Alfaro, L., Chanda, A., Kalemli-Ozcan, S., and Sayek, S. (2004). FDI and economic growth: The role of local financial markets. *Journal of International Economics*, 64(1), 89–112.

Alfaro, L., Chanda, A., Kalemli-Ozcan, S., and Sayek, S. (2010). Does foreign direct investment promote growth? Exploring the role of financial markets on linkages. *Journal of Development Economics*, 91(2), 242–256.

Alfaro, L., Kalemli-Ozcan, S., and Sayek, S. (2009). FDI, productivity and financial development. *The World Economy*, 32(1), 111–135.

Almeida, P. (1996). Knowledge sourcing by foreign multinationals: Patent citation analysis in the US semiconductor industry. *Strategic Management Journal*, 17(S2), 155–165.

Almeida, P., and Kogut, B. (1999). Localization of knowledge and the mobility of engineers in regional networks. *Management Science*, 45(7), 905–917.

Almeida, P., and Phene, A. (2004). Subsidiaries and knowledge creation: The influence of the MNC and host country on innovation. *Strategic Management Journal*, 25(8/9), 847.

Alvarez, I., and Molero, J. (2005). Technology and the generation of international knowledge spillovers: An application to Spanish manufacturing firms. *Research Policy*, 34(9), 1440–1452.

Amin, A., and Cohendet, P. (2004). *Architectures of Knowledge: Firms, Capabilities, and Communities*. Oxford: Oxford University Press.

Andersson, U., Björkman, I., and Forsgren, M. (2005). Managing subsidiary knowledge creation: The effect of control mechanisms on subsidiary local embeddedness. *International Business Review*, 14(5), 521–538.

Andersson, U., Forsgren, M., and Holm, U. (2002). The strategic impact of external networks: Subsidiary performance and competence development in the multinational corporation. *Strategic Management Journal*, 23(11), 979–996.

Andersson, U., Forsgren, M., and Holm, U. (2007). Balancing subsidiary influence in the federative MNC: A business network view. *Journal of International Business Studies*, 38(5), 802–818.

Archibugi, D., and Michie, J. (1995). The globalisation of technology: A new taxonomy, *Cambridge Journal of Economics*, 19, 121–140.

Arora, A., and Gambardella, A. (1994). The changing technology of technological change: General and abstract knowledge and the division of innovative labour. *Research Policy*, 23(5), 523–532.

Arora, A., Belenzon, S., and Rios, L.A. (2011). *The Organization of R&D in American Corporations: The Determinants and Consequences of Decentralization* NBER working paper N° 17013.

Arrow, K.J. (1962). The economic implications of learning by doing. *The Review of Economic Studies*, (29), 155–173.

Asakawa, K. (2001). Organizational tension in international R&D management: The case of Japanese firms. *Research Policy*, 30(5), 735–757.

Asheim, B., and Gertler, M. (2005). The geography of innovation. In J. Fagerberg, D.C. Mowery, and R.R. Nelson (eds), *The Oxford Handbook of Innovation*. Oxford: Oxford University Press, 292–317.

Asheim, B.T., and Isaksen, A. (2002). Regional innovation systems: The integration of local "sticky" and global "ubiquitous" knowledge. *Journal of Technology Transfer*, 27(1), 77–86.

Audretsch, D.B., and Feldman, M.P. (1996). R&D spillovers and the geography of innovation and production. *American Economic Review*, 630–640.

Barba Navaretti, G., and Venables, A.J. (2004). *Multinational Firms in the World Economy*. Princeton, NJ: Princeton University Press.

Barrios, S., Bertinelli, L., and Strobl, E. (2006). Coagglomeration and spillovers. *Regional Science and Urban Economics*, 36(4), 467–481.

Bartlett, C.A. (1979). *Multinational Structural Evolution: The Changing Decision Environment in International Divisions* (Doctoral dissertation, Harvard University, Graduate School of Business Administration).

Bartlett, C.A. (1986). Building and managing the transnational: The new organizational challenge. In M.E. Porter (ed.), *Competition in Global Industries*, Boston, MA: Harvard Business School Press, 367–401.

Bartlett, C.A., and Ghoshal, S. (1990). Managing innovation in the transnational corporation. In C.A. Bartlett, Y. Doz, and G. Hedlund (eds), *Managing the Global Firm*. Routledge: London; pp. 215–255.

Bathelt, H., Malmberg, A., and Maskell, P. (2004). Clusters and knowledge: Local buzz, global pipelines and the process of knowledge creation. *Progress in Human Geography*, 28(1), 31–56.

Belderbos, R., Capannelli, G., and Fukao, K. (2001). Backward vertical linkages of foreign manufacturing affiliates: Evidence from Japanese multinationals. *World Development*, 29(1), 189–208.

Belderbos, R., Leten, B., and Suzuki, S. (2013). How global is R&D? Firm-level determinants of home-country bias in R&D. *Journal of International Business Studies*, 44(8), 765–786.

Belderbos, R., Lykogianni, E., and Veugelers, R. (2008). Strategic R&D location in European manufacturing industries. *Review of World Economics*, 144(2), 183–206.

Ben Hamida, L. (2013). Are there regional spillovers from FDI in the Swiss manufacturing industry?. *International Business Review*, 22(4), 754–769.

Ben Hamida, L., and Gugler, P. (2009). Are there demonstration-related spillovers from FDI?: Evidence from Switzerland. *International Business Review*, 18(5), 494–508.

Berry, C.R., and Glaeser, E.L. (2005). The divergence of human capital levels across cities. *Papers in Regional Science*, 84(3), 407–444.

Beugelsdijk, S., and Mudambi, R. (2013). MNEs as border-crossing multi-location enterprises: The role of discontinuities in geographic space. *Journal of International Business Studies*, 44(5), 413–426.

Beugelsdijk, S., McCann, P., and Mudambi, R. (2010). Introduction: Place, space and organization – economic geography and the multinational enterprise. *Journal of Economic Geography*, 10(4), 485–493.

Beugelsdijk, S., Smeets, R., and Zwinkels, R. (2008). The impact of horizontal and vertical FDI on host's country economic growth. *International Business Review*, 17(4), 452–472.

Birkinshaw, J.M. (1997). Entrepreneurship in multinational corporations: The characteristics of subsidiary initiatives. *Strategic Management Journal*, 18(3), 207–229.

Birkinshaw, J.M. (2000). *Entrepreneurship in the Global Firm: Enterprise and Renewal*. London: Sage.

Birkinshaw, J.M., and Hood, N. (1998). Multinational subsidiary evolution: Capability and charter change in foreign-owned subsidiary companies. *Academy of Management Review*, 23(4), 773–795.

Birkinshaw, J.M., and Morrison, A.J. (1995). Configurations of strategy and structure in subsidiaries of multinational corporations. *Journal of International Business Studies*, 26(4), 729–753.

Birkinshaw, J.M., Hood, N., and Jonsson, S. (1998). Building firm-specific advantages in multinational corporations: The role of subsidiary initiative. *Strategic Management Journal*, 19(3), 221–242.

Blalock, G., and Gertler, P.J. (2009). How firm capabilities affect who benefits from foreign technology. *Journal of Development Economics*, 90(2), 192–199.

Blalock, G., and Simon, D.H. (2009). Do all firms benefit equally from downstream FDI? The moderating effect of local suppliers' capabilities on productivity gains. *Journal of International Business Studies*, 40(7), 1095–1112.

Blanc, H., and Sierra, C. (1999). The internationalisation of R&D by multinationals: A trade-off between external and internal proximity. *Cambridge Journal of Economics*, 23(2), 187–206.

Blomström, M., and Kokko, A. (1998). Multinational corporations and spillovers. *Journal of Economic Surveys*, 12(3), 247–277.

Blomström, M., and Sjöholm, F. (1999). Technology transfer and spillovers: Does local participation with multinationals matter?. *European Economic Review*, 43(4), 915–923.

Borensztein, E., De Gregorio, J., and Lee, J.W. (1998). How does foreign direct investment affect economic growth?. *Journal of International Economics*, 45(1), 115–135.

Boschma, R. (2005). Proximity and innovation: A critical assessment. *Regional studies*, 39(1), 61–74.

Boschma, R., and Iammarino, S. (2009). Related variety, trade linkages, and regional growth in Italy. *Economic Geography*, 85(3), 289–311.

Bouquet, C., and Birkinshaw, J. (2008). Weight versus voice: How foreign subsidiaries gain attention from corporate headquarters. *Academy of Management Journal*, 51(3), 577–601.

Branstetter, L. (2006). Is foreign direct investment a channel of knowledge spillovers? Evidence from Japan's FDI in the United States. *Journal of International Economics*, 68(2), 325–344.

Breschi, S. (2000). The geography of innovation: A cross-sector analysis. *Regional Studies*, 34(3), 213–229.

Breschi, S., and Malerba, F. (1997). Sectoral systems of innovation: Technological regimes, Schumpeterian dynamic, and spatial boundaries. In C. Edquist (ed.), *Systems of Innovation*, London: Pinter, 130–156.

Buckley, P.J., and Casson, M. (1976). *The Future of the Multinational Enterprise*. London: Palgrave Macmillan.

Buckley, P.J., Clegg, J., and Wang, C. (2002). The impact of inward FDI on the performance of Chinese manufacturing firms. *Journal of International Business Studies*, 33(4), 637–655.

Buckley, P.J., Wang, C., and Clegg, J. (2007). The impact of foreign ownership, local ownership and industry characteristics on spillover benefits from foreign direct investment in China. *International Business Review*, 16(2), 142–158.

Cantwell, J. (1989). *Technological Innovation and Multinational Corporations*. Oxford: Basil Blackwell.

Cantwell, J. (1995a). Multinational corporations and innovatory activities: Towards a new evolutionary approach. In J. Molero (ed.), *Technological Innovation, Multinational Corporations and New International Competitiveness. The Case of Intermediate Countries.* Harwood Academic Publishers Luxembourg, 21–57.

Cantwell, J. (1995b). The globalisation of technology: What remains of the product cycle model?. *Cambridge Journal of Economics*, 19, 155–174.

Cantwell. J. (2001). Innovation and information technology in MNE. In A.M. Rugman and T.L. Brewer (eds), *The Oxford Handbook of International Business*, Oxford: Oxford University Press, 343–356.

Cantwell, J. (2009). Location and the multinational enterprise. *Journal of International Business Studies*, 40(1), 35–41.

Cantwell, J., and Janne, O. (1999). Technological globalisation and innovative centres: The role of corporate technological leadership and locational hierarchy. *Research Policy*, 28(2), 119–144.

Cantwell, J., and Mudambi, R. (2005). MNE competence–creating subsidiary mandates. *Strategic Management Journal*, 26(12), 1109–1128.

Cantwell, J., and Mudambi, R. (2011). Physical attraction and the geography of knowledge sourcing in multinational enterprises. *Global Strategy Journal*, 1(3–4), 206–232.

Cantwell, J., and Piscitello, L. (2000). Accumulating technological competence: Its changing impact on corporate diversification and internationalization. *Industrial and Corporate Change*, 9(1), 21–51.

Cantwell, J., and Piscitello, L. (2005). Recent location of foreign-owned R&D activities by large multinational corporations in the European regions: The role of spillovers and externalities. *Regional Studies*, 39(1), 1–16.

Cantwell, J., and Santangelo, G.D. (1999). The frontier of international technology networks: Sourcing abroad the most highly tacit capabilities. *Information Economics and Policy*, 11(1), 101–123.

Cantwell, J., and Santangelo, G.D. (2002). The new geography of corporate research in information and communications technology (ICT). *Journal of Evolutionary Economics*, 12(1–2), 163–197.

Cantwell, J., Dunning, J.H., and Lundan, S.M. (2009). An evolutionary approach to understanding international business activity: The co-evolution of MNEs and the institutional environment. *Journal of International Business Studies*, 41(4), 567–586.

Cairncross, F. (1997). *The Death of Distance*. London: Orion Business Books.

Carlsson, B., and Stankiewicz, R. (1995). On the nature, function and composition of technological systems. In B. Carlsson (ed.), *Technological Systems and Economic Performance: The Case of Factory Automation*. Kluwer, Dordrecht, 21–56.

Castellani, D., and Zanfei, A. (2006). *Multinational Firms, Innovation and Productivity*. Cheltenham (UK): Edward Elgar.

Castellani, D., Jimenez, A., and Zanfei, A. (2013). How remote are R&D labs [quest] Distance factors and international innovative activities. *Journal of International Business Studies*, 44(7), 649–675.

Caves, R. (1982). *Multinational Enterprise and Economic Analysis.* Cambridge: Cambridge University Press.

Chang, S.J., and Xu, D. (2008). Spillovers and competition among foreign and local firms in China. *Strategic Management Journal,* 29(5), 495–518.

Chen, H., and Chen, T.J. (1998). Network linkages and location choice in foreign direct investment. *Journal of International Business Studies,* 29(3), 445–467.

Chesbrough, H. (2003). *Open Innovation: The New Imperative for Creating and Profiting from Technology.* Boston, MA: Harvard Business School Press.

Choong, C.K. (2012). Does domestic financial development enhance the linkages between foreign direct investment and economic growth?. *Empirical Economics,* 42(3), 819–834.

Chung, W. (2001). Identifying technology transfer in foreign direct investment: Influence of industry conditions and investing firm motives. *Journal of International Business Studies,* 32(2), 211–229.

Chung, W., and Alcácer, J. (2002). Knowledge seeking and location choice of foreign direct investment in the United States. *Management Science,* 48(12), 1534–1554.

Ciabuschi, F., Dellestrand, H., and Martín, O.M. (2011). Internal embeddedness, headquarters involvement, and innovation importance in multinational enterprises. *Journal of Management Studies,* 48(7), 1612–1639.

Coff, R.W. (1999). When competitive advantage doesn't lead to performance: The resource-based view and stakeholder bargaining power. *Organization Science,* 10(2), 119–133.

Cohen, W.M., and Levinthal, D.A. (1990). Absorptive capacity: A new perspective on learning and innovation. *Administrative Science Quarterly,* 128–152.

Coleman, J. (1988). Social Capital in the Creation of Human Capital. *American Journal of Sociology,* 94, 95–120.

Collins, H. (1992). *Changing Order: Replication and Induction in Scientific Practice.* Chicago: University of Chicago Press.

Collins, H.M. (2001). Tacit knowledge, trust and the Q of sapphire. *Social Studies of Science,* 31(1), 71–85.

Cooke, P., and Morgan, K. (1994). The regional innovation system in Baden-Wurttemberg. *International Journal of Technology Management,* 9(3–4), 394–429.

Coyle, D. (1997). *The Weightless World*. London: Capstone.

Crespo, N., Fontoura, M.P., and Proença, I. (2009). FDI spillovers at regional level: Evidence from Portugal. *Papers in Regional Science*, 88(3), 591–607.

Dachs, B. (2014). R&D internationalization and the global financial crisis. In B. Dachs, R. Stehrer, and G. Zahradnik (eds), *The Internationalisation of Business R&D*. Cheltenham (UK): Edward Elgar, 183–196.

Damijan, J.P., Rojec, M., Majcen, B., and Knell, M. (2013). Impact of firm heterogeneity on direct and spillover effects of FDI: Micro-evidence from ten transition countries. *Journal of Comparative Economics*, 41(3), 895–922.

de Faria, P., and Sofka, W. (2010). Knowledge protection strategies of multinational firms – A cross-country comparison. *Research Policy*, 39(7), 956–968.

De Propris, L., and Driffield, N. (2006). The importance of clusters for spillovers from foreign direct investment and technology sourcing. *Cambridge Journal of Economics*, 30(2), 277–291.

Di Minin, A., and Bianchi, M. (2011). Safe nests in global nets: Internationalization and appropriability of R&D in wireless telecom. *Journal of International Business Studies*, 42(7), 910–934.

Dörrenbächer, C., and Geppert, M. (2011). *Politics and Power in Multinational Corporation: The Role of Institutions, Interests and Identities*. Cambridge: Cambridge University Press.

Driffield, N.L. (2004). Regional policy and spillovers from FDI in the UK. *Annals of Regional Science*, 38(4), 579–594.

Driffield, N.L. (2006). On the search for spillovers from FDI with spatial dependency. *Regional Studies*, 40,107-119.

Driffield, N.L., and Love, H.J. (2007). Linking FDI motivation and host-economy productivity effects: Conceptual and empirical analysis. *Journal of International Business Studies*, 38, 460–473.

Driffield, N.L. Munday, M., and Roberts, A. (2004). Inward investment, transactions linkages, and productivity spillovers. *Papers in Regional Science*, 83, 699–722.

Dunning J.H. (1970). *Studies in International Investment*. London: Allen and Unwin.

Dunning, J.H. (1977). Trade, location of economic activity and the MNE: A search for an eclectic approach. In B. Ohlin, P. O. Hesselborn, and P. M. Wijkman (eds), *The International Allocation of Economic Activity*. London: Palgrave Macmillan, 398–418.

Dunning, J.H. (1993). *Multinational Enterprises and the Global Economy*. Wokingham: Addison-Wesley.

Dunning, J.H. (1994). Multinational enterprises and the globalization of innovatory capacity. *Research Policy*, 23(1), 67–88.

Dunning, J.H., and Lundan, S. (2008). *Multinational Enterprises and the Global Economy*, 2nd edn. Cheltenham: Edward Elgar.

Dunning J.H., and Lundan S.M. (2009). The internationalization of Corporate R&D: A review of the evidence and some policy implications for home countries, *Review of Policy Research*, 26, 13–33.

Eapen, A. (2012). Social structure and technology spillovers from foreign to domestic firms. *Journal of International Business Studies*, 43, 244–263.

Eden, L. (2009). Letter from the editor-in-chief: FDI spillovers and linkages. *Journal of International Business Studies*, 40, 1065–1069.

Estall, R.C., and Buchanan, R.O. (1961). *Industrial Activity and Economic Geography*. London: Hutchinson.

Ethier, W.J., and Markusen, J.R. (1996). Multinational firms, technology diffusion and trade. *Journal of International Economics*, 41, 1–28.

Ferner, A., Almond, P., and Cooling, T. (2005). Institutional theory and the cross-national transfer of employment policy: The case of "Workforce Diversity" in US multinationals. *Journal of International Business Studies*, 36, 304–322.

Florida, R. (1997). The globalization of R&D: Results of a survey of foreign-affiliated R&D laboratories in the USA. *Research Policy*, 26(1), 85–103.

Florida, R. (2005). The world is spiky. *Atlantic Monthly* October, 48–51.

Findlay, R. (1978). Relative backwardness, direct foreign investment, and the transfer of technology: A simple dynamic model. *Quarterly Journal of Economics*, 92, 1–16.

Ford, T.C., Rork, J.C., and Elmslie, B.T. (2008). Considering the source: Does the country of origin of FDI matter to economic growth? *Journal of Regional Science*, 48, 329–357.

Fors, G. (1996). *R&D and Technology Transfer by Multinational Enterprises*. Stockholm: Almquist and Wiksell and IUI.

Forsgren, M., Holm, U., and Johanson, J. (1992). Internationalization of the second degree: The emergence of European-based centres in Swedish firms. In S. Young and J. Hamill (eds), *Europe and the Multinationals*. Aldershot: Edward Elgar, 235–253.

Fosfuri, A., Motta, M., and Rønde, T. (2001). Foreign direct investment and spillovers through workers' mobility. *Journal of International Economics*, 53, 205–222.

Foss, N.J. (1997). *Resources, Firms, and Strategies: A Reader in the Resource-Based Perspective*. Oxford: Oxford University Press.

Freeman, C. (1987). *Technology Policy and Economic Performance: Lessons from Japan*. London: Frances Pinter.

Freeman, C., and Hagerdoon, J. (1995). Convergence and divergence in the internationalization of technology. In J. Hagerdoon (ed.), *Technical Change and the World Economy: Convergence and Divergence in Technology Strategies*. Aldershot: Edward Elgar, 34–57.

Friedman, T. (2005). *The World Is Flat: A Brief History of the Twenty-First Century*. New York: Farrar, Straus, and Giroux.

Frost, T.S. (2001). The geographic sources of foreign subsidiaries' innovations. *Strategic Management Journal*, 22(2), 101–124.

Frost, T.S., Birkinshaw, J.M., and Ensign, P.C. (2002). Centers of excellence in multinational corporations. *Strategic Management Journal*, 23(11), 997–1018.

Gaspar, J., and Glaeser, E.L. (1998). Information technology and the future of cities. *Journal of Urban Economics*, 43(1), 136–156.

Gassmann, O., and von Zedwitz, M. (1999). New concepts and trends in international R&D organization. *Research Policy*, 28, 231–250.

Gates, S.R., and Egelhoff, W.G. (1986). Centralization in headquarters-subsidiary relationships. *Journal of International Business Studies*, 17(2), 71–92.

Gertler, M.S. (2001). Best practice? Geography, learning and the institutional limits to strong convergence. *Journal of Economic Geography*, 1(1), 5–26.

Gertler, M.S. (2003). Tacit knowledge and the economic geography of context, or the undefinable tacitness of being (there). *Journal of Economic Geography*, 3(1), 75–99.

Gertler, M.S., and Levitte, Y.M. (2005). Local nodes in global networks: The geography of knowledge flows in biotechnology innovation. *Industry and Innovation*, 12(4), 487–507.

Gerybadze, A., and Reger, G. (1999). Globalization of R&D: Recent changes in the management of innovation in transnational corporations. *Research policy*, 28(2), 251–274.

Ghemawat, P. (2001). Distance still matters. *Harvard Business Review*, 79(8), 137–147.

Ghoshal, S. (1986). *The Innovative Multinational: A Differentiated Network of Organizational Roles and Management Processes: A Thesis*. Harvard University, Graduate School of Business Administration.

Ghoshal, S., and Bartlett, C.A. (1988). Creation, adoption, and diffusion of innovations by subsidiaries of multinational corporations. *Journal of International Business Studies*, 19(3), 365–388.

Ghoshal, S., and Bartlett, C.A. (1990). The multinational corporation as an interorganizational network. *Academy of Management Review*, 15(4), 603–626.

Girma, S. (2005). Absorptive capacity and productivity spillovers from fdi: A threshold regression analysis. *Oxford Bulletin of Economics and Statistics*, 67(3), 281–306.

Girma, S., Gong, Y., and Holger, G. (2008). Foreign direct investment, access to finance, and innovation activity in Chinese enterprises. *The World Bank Economic Review*, 22, 367–382.

Giroud, A. (2007). MNEs vertical linkages: The experience of Vietnam after Malaysia. *International Business Review*, 16(2), 159–176.

Giroud. A., and Scott-Kennel, J. (2009). MNE linkages in International Business: A framework for analysis. *International Business Review*, 18(6), 555–566.

Giroud, A., Jindra, B., and Marek, P. (2012). Heterogeneous FDI in transition economies – A novel approach to assess the developmental impact of backward linkages. *World Development*, 40, 2206–2220.

Giuliani, E. (2007). The selective nature of knowledge networks in clusters: Evidence from the wine industry. *Journal of Economic Geography*, 7(2), 139–168.

Giuliani, E., and Bell, M. (2005). The micro-determinants of meso-level learning and innovation: Evidence from a Chilean wine cluster. *Research Policy*, 34(1), 47–68.

Glaeser, E., Kallal, H., Scheinkman, J., and Shleifer, A. (1992). Growth in Cities. *Journal of Political Economy*, 100(6), 1126–1152.

Glass, A.J., and Saggi, K. (1998). International technology transfer and the technology gap. *Journal of Development Economics*, 55, 369–398.

Glass, A.J. and Saggi, K. (2002). Multinational Firms and Technology Transfer. *Scandinavian Journal of Economics*, 104, 495–514.

Goerzen, A., Asmussen, C.G., and Nielsen, B.B. (2013). Global cities and multinational enterprise location strategy. *Journal of International Business Studies*, 44(5), 427–450.

Gordon, I.R., and McCann, P. (2000). Industrial clusters: Complexes, agglomeration and/or social networks?. *Urban studies*, 37(3), 513–532.

Görg, H., and Strobl, E. (2001). Multinational companies and productivity spillovers: A meta-analysis. *Economic Journal*, 111, 723–739.

Graham, E.M. (1992). *Japanese Control of R&D Activities in the United States: Is This Cause for Concern? Japan's Growing Technological Capability*. Washington, DC: National Academy Press.

Granovetter, M.S. (1973). The strength of weak ties. *American Journal of Sociology*, 78(6), 1360–1380.

Granstrand, O. (1999). Internationalization of corporate R&D: A study of Japanese and Swedish corporations. *Research Policy*, 28(2), 275–302.

Granstrand, O., Håkanson, L., and Sjölander, S. (1993). Internationalization of R&D – a survey of some recent research. *Research Policy*, 22(5), 413–430.

Gulati, R. (1995). Social structure and alliance formation patterns: A longitudinal analysis. *Administrative Science Quarterly*, 40(4), 619–652.

Gupta, A.K., and Govindarajan, V. (2000). Knowledge flows within multinational corporations. *Strategic Management Journal*, 21(4), 473–496.

Haddad, M., and Harrison, A. (1993). Are there positive spillovers from direct foreign investment?: Evidence from panel data for Morocco. *Journal of Development Economics*, 42, 51–74.

Hakanson, L. (1990) International Decentralization of R&D – The Organizational Challenges, In: Bartlett, C.A., Doz, Y. and Hedlund, G. (Eds.): *Managing the Global Firm*, London, Routledge.

Håkanson, L. (1992). Locational determinants of foreign R&D in Swedish multinationals. In O. Granstrand, L. Håkanson, and S. Sjolander (eds), *Technology Management and International Business: Internationalisation of R&D and Technology*. Chichester, UK: Wiley.

Håkanson, L., and Nobel, R. (1993). Foreign research and development in Swedish multinationals. *Research Policy*, 22(5), 373–396.

Hannan, M.T., and Freeman, J. (1977). The population ecology of organizations. *American Journal of Sociology*, 82(5), 929–964.

Hannigan, T.J., Cano-Kollmann, M., and Mudambi, R. (2015). Thriving innovation amidst manufacturing decline: The Detroit auto cluster and the resilience of local knowledge production, *Industrial and Corporate Change*, 24(3), 613–634.

Hansen, M.T. (1999). The search-transfer problem: The role of weak ties in sharing knowledge across organization subunits. *Administrative Science Quarterly*, 44(1), 82–111.

Haskel, J.E., Pereira, S.C., and Slaughter, M.J. (2007). Does inward foreign direct investment boost the productivity of domestic firms? *Review of Economics and Statistics*, 89, 482–496.

Hedlund, G. (1986). The hypermodern MNC – A heterarchy?. *Human Resource Management*, 25(1), 9–35.

Henderson, R., and Cockburn, I. (1996). Scale, scope, and spillovers: The determinants of research productivity in drug discovery. *The Rand Journal of Economics*, 27(1), 32–59.

Hennart, J.F. (1982). *A Theory of Multinational Enterprise*. Ann Arbor, MI: University of Michigan Press.

Hennart, J.F., and Larimo, J. (1998). The impact of culture on the strategy of multinational enterprises: Does national origin affect ownership decisions?. *Journal of International Business Studies*, 29(3), 515–538.

Hewitt, G. (1980). Research and development performed abroad by US manufacturing multinationals. *Kyklos*, 33(2), 308–327.

Hirschey, R., and Caves, R. (1981). Research and transfer of technology by multinational enterprises. *Oxford Bulletin of Economics and Statistics*, 43(2): 115–130.

Hitt, M.A., Beamish, P.W., Jackson, S.E., and Mathieu, J.E. (2007). Building theoretical and empirical bridges across levels: Multilevel research in management. *Academy of Management Journal*, 50(6), 1385–1399.

Holm, U., and Pedersen T. (2000). *The Emergence and Impact of MNC Centres of Excellence*. London: Palgrave Macmillan.

Hong, E., and Sun, L. (2011). Foreign direct investment and total factor productivity in China: A spatial dynamic panel analysis. *Oxford Bulletin of Economics and Statistics*, 73, 771–791.

Hood, N., and Young, S. (1976). US investment in Scotland: Aspects of the branch factory syndrome. *Scottish Journal of Political Economy*, 23, 279–294.

Hood, N. and Young, S. (1988). Inward investment and the EEC: UK evidence on corporate integration strategies. In J.H. Dunning and P. Robson (eds), *Multinationals and the European Community*, Oxford: Blackwell, 91–104.

Hood, N., Young, S., and Lal, D. (1994). Strategic evolution within Japanese manufacturing plants in Europe. UK evidence. *International Business Review*, 3(2), 97–122.

Hoover, E.M. (1937). *Location Theory and the Shoe Leather Industries* (Vol. 55). Cambridge: Harvard University Press.

Howells, J. (1999). Regional systems of innovation? In D. Archibugi & J. Michie (Eds.), *Innovation Policy in a Global Economy* (pp. 67–93). Cambridge: Cambridge University Press.

Huang, L., Liu, X., and Xu, L. (2012). Regional innovation and spillover effects of foreign direct investment in China: A threshold approach. *Regional Studies*, 46(5), 583–596.

Hymer, S.H. (1976). *The International Operations of National Firms: A Study of Direct Foreign Investment*. Cambridge, MA: MIT press.

Iammarino, S., and McCann, P. (2013). *Multinationals and Economic Geography: Location, Technology and Innovation*. Cheltenham (UK): Edward Elgar.

IFI Claims. (2015). *IFI Claims 2014 Top 50 US Patent Assignees*, http://www.ificlaims.com/index.php?page=misc_top_50_2014, accessed January 01, 2015

Iwasa, T., and Odagiri, H. (2004). Overseas R&D, knowledge sourcing and patenting: An empirical study of Japanese R&D investment in the US. *Research Policy*, 33(5), 807–828.

Jacobs, J. (1961). *The Death and Life of Great American Cities*. London: Vintage.

Jacobs, J. (1969). *The Economy of Cities*. New York: Random House.

Jaffe, A., Trajtenberg, M., and Henderson, R. (1993). Geographic localization of knowledge spillovers as evidenced by patent citations. *Quarterly Journal of Economics*, 108(3), 577–598.

Jarillo, J.C., and Martínez, J.I. (1990). Different roles for subsidiaries: The case of multinational corporations in Spain. *Strategic Management Journal*, (11)7, 501–513.

Jaruzelski B., Loehr J., Holman R. (2011). *The Global Innovation 1000. Why Culture Is Key*, 65, Strategy+Business, http://www.strategyand.pwc.com, accessed on March 15, 2015.

Jaruzelski B., Staack V., Goehle B. (2014). *The Global Innovation 1000. Proven Paths to Innovation Success*, 77, Strategy+Business, http://www.strategyand.pwc.com, accessed on March 15, 2015.

Javorcik, B.S. (2004). Does Foreign Direct Investment increase the productivity of domestic firms? In search of spillovers through backward linkages. *American Economic Review*, 24(2), 605–626.

Javorcik, B.S., and Spatareanu, M. (2008). To share or not to share: Does local participation matter for spillovers from foreign direct investment? *Journal of Development Economics*, 85, 194–217.

Javorcik, B.S. and Spatareanu, M. (2011). Does it matter where you come from? Vertical spillovers from foreign direct investment and the origin of investors. *Journal of Development Economics*, 96, 126–138.

Jindra, B., Giroud, A., and Scott-Kennel, J. (2009). Subsidiary roles, vertical linkages and economic development: Lessons from transition economies. *Journal of World Business*, 44, 167–179.

Johanson, J., and Vahlne, J.E. (2009). The Uppsala internationalization process model revisited: From liability of foreignness to liability of outsidership. *Journal of International Business Studies*, 40(9), 1411–1431.

Jordaan, J.A. (2005). Determinants of FDI-induced externalities: New empirical evidence for Mexican manufacturing industries. *World Development*, 33, 2103–2118.

Katila R. (2002). New product search over time: Past ideas in their prime? *Academy of Management Journal*, 45, 995 – 1010.

Kemeny, T. (2010). Does foreign direct investment drive technological upgrading? *World Development*, 38, 1543–1554.

Kindleberger, C.P. (1969). *American Business Abroad: Six Lectures on Direct Investment*. New Haven, CT: Yale University Press.

Kogut, B., and Zander, U. (1993). Knowledge of the firm and the evolutionary theory of the multinational corporation. *Journal of International Business Studies*, 24(4), 625–645.

Kohpaiboon, A. (2006). Foreign direct investment and technology spillover: A cross-industry analysis of Thai manufacturing. *World Development*, 34, 541–556.

Kokko, A. (1994). Technology, market characteristics, and spillovers. *Journal of Development Economics*, 43, 279–293.

Kristensen, P.H., and Zeitlin, J. (2005). *Local Players in Global Games: The Strategic Constitution of a Multinational Corporation*. Oxford: Oxford University Press.

Kuemmerle, W. (1997). Building effective R&D capabilities abroad. *Harvard Business Review*, 75(2), 61–70.

Kuemmerle, W. (1998). Optimal scale for research and development in foreign environments: An investigation into size and performance of research and development laboratories abroad. *Research Policy*, 27(2), 111–126.

Kuemmerle, W. (1999). The drivers of foreign direct investments into research and development – an empirical investigation, *Journal of International Business Studies*, 30, 1–24.

Kumar, N. (1996). Intellectual property protection, market orientation and location of overseas R&D activities by multinational enterprises. *World Development*, 24, 673–688.

Kumar, N. (2001). Determinants of location of overseas R&D activity of multinational enterprises: The case of US and Japanese corporations. *Research Policy*, 30, 159–174.

Kumaraswamy, A., Mudambi, R., Saranga, H., and Tripathy, A. (2012). Catch-up strategies in the Indian auto components industry: Domestic firms' responses to market liberalization. *Journal of International Business Studies*, 43, 368–395.

Lahiri, N. (2010). Geographic distribution of R&D activity: How does it affect innovation quality?. *Academy of Management Journal*, 53(5), 1194–1209.

Laursen, K., and Salter, A. (2006). Open for innovation: The role of openness in explaining innovation performance among UK manufacturing firms. *Strategic Management Journal*, 27(2), 131–150.

Le Bas, C., and Sierra, C. (2002). "Location versus home country advantages" in R&D activities: Some further results on multinationals' locational strategies. *Research Policy*, 31(4), 589–609.

Leonard-Barton, D. (1995). *Wellspring of knowledge*. Boston, MA: Harvard Business School Press.

Lewin, A.Y., Massini, S., and Peeters, C. (2009). Why are companies offshoring innovation? The emerging global race for talent. *Journal of International Business Studies*, 40(6), 901–925.

Li, X., and Liu, X. (2005). Foreign direct investment and economic growth: An increasingly endogenous relationship. *World Development*, 33, 393–407.

Liu, X., and Buck, T. (2007). Innovation performance and channels for international technology spillovers: Evidence from Chinese high-tech industries. *Research Policy*, 36, 355–366.

Liu, X., and Zhou, H. (2008). The impact of Greenfield FDI and mergers and acquisitions on innovation in Chinese high-tech industries. *Journal of World Business*, 43, 352–364.

Liu, X., Wang, C., and Wei, Y. (2009). Do local manufacturing firms benefit from transactional linkages with multinational enterprises in China? *Journal of International Business Studies*, 40, 1113–1130.

Liu, X., Siler, P., Wang, C., and Wei, Y. (2000). productivity spillovers from foreign direct investment: Evidence from UK Industry level panel data. *Journal of International Business Studies*, 31, 407–425.

Lorenzen, M., and Mudambi, R. (2013). Clusters, connectivity and catch-up: Bollywood and Bangalore in the global economy. *Journal of Economic Geography*, 13(3), 501–534.

Lundvall, B.Å. (1992). *National Systems of Innovation: Towards a Theory of Innovation and Interactive Learning*. London: Pinter.

Lundvall, B.Å. (2007). National innovation systems – analytical concept and development tool. *Industry and Innovation*, 14(1), 95–119.

Malmberg, A., and Maskell, P. (2002). The elusive concept of localization economies: Towards a knowledge-based theory of spatial clustering. *Environment and Planning A*, 34(3), 429–450.

Mansfield, E., Teece, D., and Romeo, A. (1979). Overseas research and development by U.S.-based firms. *Economica*, 46, 187–196.

March, J.G. (1991). Exploration and exploitation in organizational learning. *Organization Science*, 2(1), 71–87.

Marin, A., and Bell, M. (2006). Technology spillovers from Foreign Direct Investment: The active role of MNC subsidiaries in Argentina in the 1990. *Journal of Development Studies*, 42, 678–697.

Marin, A., and Sasidharan, S. (2010). Heterogeneous MNC subsidiaries and technological spillovers: Explaining positive and negative effects in India. *Research Policy*, 39, 1227–1241.

Mariotti, S., and Piscitello, L. (1995). Information costs and location of FDIs within the host country: Empirical evidence from Italy. *Journal of International Business Studies*, 26, 815–841.

Mariotti, S., Piscitello, L., and Elia, S. (2010). Spatial agglomeration of multinational enterprises: The role of information externalities and knowledge spillovers. *Journal of Economic Geography*, 10(4), 519–538.

Markusen, A. (1996). Sticky places in slippery space: A typology of industrial districts. *Economic Geography*, 72, 293–313.

Marshall. A. (1890), *Principles of Economics*. London: Palgrave Macmillan.

Marshall, A.,. (1920). *Principles of Economics*. London: Palgrave Macmillan.

Martin, R., and Sunley, P. (2003). Deconstructing clusters: Chaotic concept or policy panacea? *Journal of Economic Geography*, 3(1), 5–35.

Maskell, P., and Malmberg, A. (1999). Localised learning and industrial competitiveness. *Cambridge Journal of Economics*, 23(2), 167–185.

Maskell, P., and Malmberg, A. (2007). Myopia, knowledge development and cluster evolution. *Journal of Economic Geography*, 7(5), 603–618.

McCann, P. (2011). International business and economic geography: Knowledge, time and transaction costs. *Journal of Economic Geography*, 11(2), 309–317.

McCann, P., and Mudambi, R. (2005). Analytical differences in the economics of geography: The case of the multinational firm. *Environment and Planning A*, 37(10), 1857.

McCann, P., and Shefer, D. (2004). Location, agglomeration and infrastructure. *Papers in Regional Science*, 83(1), 177–196.

Menghinello, S., De Propis, L., and Driffield, N. (2010). Industrial districts, inward foreign investment and regional development. *Journal of Economic Geography*, 10, 539–558.

Meyer, K.E. (2004). Perspectives on multinational enterprises in emerging economies. *Journal of International Business Studies*, 35, 259–276.

Meyer, K.E., and Sinani, E. (2009). When and where does foreign direct investment generate positive spillovers? A meta-analysis. *Journal of International Business Studies*, 40, 1075–1094.

Michailova, S., and Hutchings, K. (2006). National cultural influences on knowledge sharing: A comparison of China and Russia. *Journal of Management Studies*, 43(3), 383–405.

Morgan, K. (2004). The exaggerated death of geography: Learning, proximity and territorial innovation systems. *Journal of Economic Geography*, 4(1), 3–21.

Morrissey, O. (2012). FDI in Sub-Saharan Africa: Fewer linkages, fewer spillovers. *European Journal of Development Research*, 24, 26–31.

Mowery DC, and Nelson RR. (1999). Explaining industrial leadership. In *Sources of Industrial Leadership*, Mowery DC, Nelson RR (eds). Cambridge University Press: Cambridge, MA; 359–382.

Mowery, D.C., and Rosenberg, N. (1979). The influence of market demand upon innovation: A critical review of some recent empirical studies. *Research Policy*, 8(2), 102–153.

Mudambi, R. (2008). Location, control and innovation in knowledge-intensive industries. *Journal of Economic Geography*, 10, 699–725.

Mudambi, R. (2011). Hierarchy, coordination and innovation in the multinational enterprise. *Global Strategy Journal*, 1(3–4), 317–323.

Mudambi, R., and Navarra, P. (2004). Is knowledge power? Knowledge flows, subsidiary power and rent-seeking within MNCs. *Journal of International Business Studies*, 35(5), 385–406.

Mudambi, R., and Santangelo, G.D. (2015). From shallow resource pools to emerging clusters: The role of multinational enterprise subsidiaries in peripheral areas. *Regional Studies*, (DOI:10.1080/00343404.2014.98 5199), 1–15.

Mudambi, R., Pedersen, T., and Andersson, U. (2014). How subsidiaries gain power in multinational corporations. *Journal of World Business*, 49(1), 101–113.

Mudambi, R., Piscitello, L., and Rabbiosi, L. (2014). Reverse knowledge transfer in MNEs: Subsidiary innovativeness and entry modes. *Long Range Planning*, 47(1–2), 49–63.

Narula, R. (2002). Innovation systems and "inertia" in R&D location: Norwegian firms and the role of systemic lock-in. *Research Policy*, 31(5), 795–816.

Narula, R., and Driffield, N. (2012). Does FDI cause development? The ambiguity of the evidence and why it matters. *European Journal of Development Research*, 24, 1–7.

Narula, R., and Santangelo, G.D. (2012). Location and collocation advantages in international innovation. *Multinational Business Review*, 20(1), 6–25.

Nelson, R.R. (1993). *National Innovation Systems: A Comparative Analysis*. Oxford: Oxford University Press.

Nelson, R.R., and Winter, S.G. (1982). *An Evolutionary Theory of Economic Change*. Cambridge: Harvard Business School Press.

Niosi, J., and Godin, B. (1999). Canadian R&D abroad management practices. *Research Policy*, 28(2/3), 215–230.

Nobel, R., and Birkinshaw, J. (1998). Innovation in multinational corporations: Control and communication patterns in international R&D operations. *Strategic Management Journal*, 19(5) 479–496.

Nonaka, I., and Takeuchi, H. (1995). *The Knowledge Creating Company*. Oxford: Oxford University Press.

Patel, P. (1995). Localised production of technology for global markets. *Cambridge Journal of Economics*, 19, 141–153.

Patel, P., and Pavitt, K. (1991). Large firms in the production of the world's technology: An important case of "non-globalisation". *Journal of International Business Studies*, 22(1), 1–21.

Patel, P., and Pavitt, K. (1994). The continuing, widespread (and neglected) importance of improvements in mechanical technologies. *Research Policy*, 23(5), 533–545.

Patel, P., and Vega, M. (1999). Patterns of internationalisation of corporate technology: Location vs. home country advantages. *Research Policy*, 28(2), 145–155.

Patibandla, M., and Petersen, B. (2002). Role of transnational corporations in the evolution of a high-tech industry: The case of India's software industry. *World Development*, 30(9), 1561–1577.

Pearce, R.D. (1989). *The Internationalization of Research and Development by Multinational Enterprises*. New York: St Martin's Press.

Pearce, R.D. (1999). Decentralised R&D and strategic competitiveness: Globalised approaches to generation and use of technology in multinational enterprises (MNEs). *Research Policy*, 28(2), 157–178.

Pearce, R.D., and Singh, S. (1992). Internationalisation of R&D among the world's leading enterprises. In O. Granstrand, S. Sjölander, and L. Håkanson (eds), *Technology, Management and International Business: Internationalisation of R&D and Technology*. Chichester, UK: Wiley, 137–162.

Perez, T. (1997). Multinational enterprises and technological spillovers: An evolutionary model. *Journal of Evolutionary Economics*, 7, 169–192.

Perlmutter, H.V. (1969). The tortuous evolution of the multinational corporation. *Columbia Journal of World Business*, 10(1), 9–18.

Perri, A., and Andersson, U. (2014). Knowledge outflows from foreign subsidiaries and the tension between knowledge creation and knowledge protection: Evidence from the semiconductor industry. *International Business Review*, 23, 63–75.

Perri, A., and Peruffo, E. (2015). Knowledge spillovers from FDI: A critical review from the international business perspective. *International Journal of Management Reviews*, DOI: 10.1111/ijmr.12054.

Perri, A., Andersson, U., Nell, P.C., and Santangelo, G.D. (2013). Balancing the trade-off between learning prospects and spillover risks: MNCs' subsidiaries linkage patterns in developed countries. *Journal of World Business*, 48, 503–514.

Pfeffer, J., and Salancik, G.R. (1978). *The External Control of Organizations*. New York: Harper and Row.

Phene, A., and Almeida, P. (2008). Innovation in multinational subsidiaries: The role of knowledge assimilation and subsidiary capabilities. *Journal of International Business Studies*, 39, 901–919.

Polanyi, M. (1966). *The Tacit Dimension*. Gloucester, MA: Peter Smith.

Porter, M.E. (1986). *Competition in Global Industries*. Boston, MA: Harvard Business School Press.

Porter, M.E. (1990). *The Competitive Advantage of the Nations*. New York: Free Press.

Porter, M.E. (2000). Location, competition, and economic development: Local clusters in a global economy. *Economic Development Quarterly*, 14(1), 15–34.

Potter, J., Moore, B., and Spires, R. (2002). The wider effects of inward foreign direct investment in manufacturing on UK industry. *Journal of Economic Geography*, 2(3), 279–310.

Prahalad, C.K., and Doz, Y.L. (1981). An approach to strategic control in MNCs. *Sloan Management Review*, 22(4), 5–13.

Rabbiosi, L. (2011). Subsidiary roles and reverse knowledge transfer: An investigation of the effects of coordination mechanisms. *Journal of International Management*, 17(2), 97–113.

Rabbiosi, L., and Santangelo, G.D. (2013). Parent company benefits from reverse knowledge transfer: The role of the liability of newness in MNEs. *Journal of World Business*, 48(1), 160–170.

Reagans, R., and McEvily, B. (2003). Network structure and knowledge transfer: The effects of cohesion and range. *Administrative Science Quarterly*, 48(2), 240–267.

Rodriguez-Clare, A. (1996). Multinationals, Linkages and Economic Development. *American Economic Review*, 86, 852–873.

Rogers, E., and Larsen J. (1984). *Silicon Valley Fever*. New York: Basic Books.

Romer, P.M. (1986). Increasing returns and long-run growth. *Journal of Political Economy*, 94(5), 1002–1037.

Ronstadt, R. (1978). International R&D: The establishment and evolution of research and development abroad by seven U.S. multinationals. *Journal of International Business Studies*, 9, 7–24.

Rousseau, D. (1985). Issues of level in organizational research: Multilevel and cross-level perspectives. In L.L. Cummings and B.M. Staw (eds), *Research in Organizational Behavior*. Greenwich, 1–37, CT: JAI Press.

Rugman, A. (1981). Research and development by multinational and domestic firms in Canada. *Canadian Public Policy*, 7(4), 604–616.

Sanna-Randaccio, F., and Veugelers, R. (2007). Multinational knowledge spillovers with decentralised R&D: A game-theoretic approach. *Journal of International Business Studies*, 38(1), 47–63.

Santangelo, G.D. (2009). MNCs and linkages creation: Evidence from a peripheral area. *Journal of World Business*, 44(2), 192–205.

Scalera V.G., Mukherjee D., Perri A., and Mudambi R. (2014). A longitudinal study of MNE innovation: The case of Goodyear, *Multinational Business Review*, 22(3), 270–293.

Scalera, V.G., Perri A., and Mudambi, R. (2015). Managing Innovation in Emerging Economies: Organizational Arrangements and Resources of Foreign MNEs in the Chinese Pharmaceutical Industry. In L. Tihanyi, E.R. Banalieva, T.M. Devinney, and T. Pedersen (eds), Emerging Economies and Multinational Enterprises, *Advances in International Management* (28), 201–233, Emerald Group.

Schrader, S. (1991). Informal technology transfer between firms: Cooperation through information trading. *Research Policy*, 20(2), 153–170.

Scott-Kennel, J. (2007). Foreign direct investment and local linkages: An empirical investigation. *Management International Review*, 47(1), 1–27.

Shane, S. (2001). Technology opportunity and firm formation. *Management Science*, 47(2), 205–220.

Shaver, J.M., and Flyer, F. (2000). Agglomeration economies, firm heterogeneity, and foreign direct investment in the United States. *Strategic Management Journal*, 21(12), 1175–1194.

Shimizutani, S., and Todo, Y. (2008). What determines overseas R&D activities? The case of Japanese multinational firms. *Research Policy*, 37(3), 530–544.

Simonin, B.L. (1997). The importance of collaborative know-how: An empirical test of the learning organization. *Academy of Management Journal*, 40(5), 1150–1174.

Singh, J. (2005). Collaborative networks as determinants of knowledge diffusion patterns. *Management Science*, 51(5), 756–770.

Singh, J. (2007). Asymmetry of knowledge spillovers between MNCs and host country firms. *Journal of International Business Studies*, 38, 764–786.

Singh, J. (2008). Distributed R&D, cross-regional knowledge integration and quality of innovative output. *Research Policy*, 37(1), 77–96.

Sosa, M.E., Eppinger, S.D., Pich, M., McKendrick, D.G., and Stout, S.K. (2002). Factors that influence technical communication in distributed product development: An empirical study in the telecommunications industry. *Engineering Management, IEEE Transactions on*, 49(1), 45–58.

Spencer, W.J. (2008). The impact of multinational enterprise strategy on indigenous enterprises: Horizontal spillovers and crowding out in developing countries. *Academy of Management Review*, 33, 341–361.

Stopford, J.M., and Wells, L.T. (1972). *Managing the Multinational Enterprise: Organization of the Firm and Ownership of the Subsidiaries* (Vol. 2). New York: Basic Books.

Storper, M., and Venables, A.J. (2004). Buzz: Face-to-face contact and the urban economy. *Journal of Economic Geography, 4*(4), 351–370.

Takii, S. (2005). Productivity spillovers and characteristics of foreign multinational plants in Indonesian manufacturing 1990–1995. *Journal of Development Economics, 76*, 521–542.

Teece, D. (1986). Profiting from technological innovation: Implications for integration, collaboration, licensing, and public policy. *Research Policy, 15*(6), 285–305.

Teece, D.J. (2000). Strategies for managing knowledge assets: The role of firm structure and industrial context. *Long Range Planning, 33*(1), 35–54.

Thompson, E.R. (2002). Clustering of foreign direct investment and enhanced technology transfer: Evidence from Hong Kong garment firms in China. *World Development, 30*, 873–889.

Thursby, J., and Thursby, M. (2006). *Here or There? A Survey of Factors in Multinational R&D Location. Report to the Government-University-Industry Research Roundtable.* Washington DC: National Academy of Sciences.

Tian, X. (2010). Managing FDI technology spillovers: A challenge to TNCs in emerging markets. *Journal of World Business, 45*, 276–284.

Tödtling, F., and Trippl, M. (2005). One size fits all?: Towards a differentiated regional innovation policy approach. *Research Policy, 34*(8), 1203–1219.

Trippl, M., Tödtling, F., and Lengauer L. (2009). Knowledge sourcing beyond buzz and pipelines: Evidence from the Vienna software sector, *Economic Geography, 85*, 443–462.

UNCTAD. (2005). *World Investment Report 2005.* New York: United Nations.

Uzzi, B. (1997). Social structure and competition in interfirm networks: The paradox of embeddedness. *Administrative Science Quarterly, 42*(1), 35–67.

Vernon, R. (1966). International investment and international trade in the product cycle. *Quarterly Journal of Economics, 80*, 190–207.

Vernon, R. (1979). The product cycle hypothesis in a new international environment. *Oxford Bulletin of Economics and Statistics, 41*(4), 255–267.

Veugelers, R., and Cassiman, B. (2004). Foreign subsidiaries as a channel of international technology diffusion: Some direct firm level evidence from Belgium. *European Economic Review, 48*, 455–476.

von Hippel, E. (1987). Cooperation between rivals: Informal know-how trading. *Research Policy*, 16, 291–302.

von Zedtwitz, M., and Gassmann, O. (2002). Market versus technology drive in R&D internationalisation: Four patterns of managing research and development. *Research Policy*, 31, 569–588.

von Zedtwitz, M., Gassmann, O., and Boutellier, R. (2004). Organizing global R&D: Challenges and dilemmas. *Journal of International Management*, 10(1), 21–49.

Wang, J., and Blomström, M. (1992). Foreign investment and technology transfer: A simple model. *European Economic Review*, 36, 137–155.

Watts, H.D. (1981). *The Branch Plant Economy*. London: Longman.

Wei, Y., and Liu, X. (2006). Productivity, spillovers from R&D, exports and FDI in China's manufacturing sector. *Journal of International Business Studies*, 37, 544–557.

White, R.E., and Poynter, T.A. (1984). Strategies for foreign-owned subsidiaries in Canada. *Business Quarterly*, 49(2), 59–69.

White, R.E., and Poynter, T.A. (1990). Organizing for Worldwide Advantage. In C.A. Bartlett, Y.L. Doz, and G. Hedlund (eds), *Managing the Global Firm*. London: Routledge, 95–113.

Wolfe, D.A., and Gertler, M.S. (2004). Clusters from the inside and out: Local dynamics and global linkages. *Urban Studies*, 41(5–6), 1071–1093.

Young, S., Hood, N., and Dunlop, S. (1988). Global strategies, multinational subsidiary roles and economic impact in Scotland. *Regional Studies*, 22, 487–497.

Young, S., Hood, N., and Peters, E. (1994). Multinational enterprises and regional economic development. *Regional Studies*, 28, 657–677.

Zander, I. (1997). Technological diversification in the multinational corporation – historical evolution and future prospects. *Research Policy*, 26(2), 209–227.

Zanfei, A. (2000). Transnational firms and the changing organisation of innovative activities. *Cambridge Journal of Economics*, 24(5), 515–542.

Zejan, M.C. (1990). R&D activities in affiliates of Swedish multinational enterprises. *Scandinavian Journal of Economics*, 92, 487–500.

Zhang, Y., Haiyang, L., Li, Y., and Zhou L. (2010). FDI spillovers in an emerging market: The role of foreign firms' country origin diversity and domestic firms' absorptive Capacity. *Strategic Management Journal*, 31, 969–989.

Zhao, M. (2006). Conducting R&D in countries with weak intellectual property rights protection. *Management Science*, 52(8), 1185–1199.

Index

Alcácer, J., 49, 83, 102, 116
agglomeration, 90, 103
autonomy, 54–60, 66, 77, 119

Birkinshaw, J., 43, 54–55, 65–67, 69, 73–75
border effects, 89
branch-plant syndrome, 77

Cantwell, J., 11, 15–16, 22–23, 48, 70–71, 72, 103
centripetal and centrifugal forces, 35–40
cities, 95, 98–99
clusters, 94–99, 101–104, 115
codified knowledge, 17, 37
co-location, 19–21, 101, 102, 117
competition effects, 112
connectivity, 96, 116
control, 54–60, 73
coordination costs, 36, 56

death of distance view, 17
demand-driven forces, 16, 38–39, 40–45
demonstration effects, 112
distance effects, 89
double network structure, 16, 72

eclectic paradigm, 45, 87, 113
economic geography, 87–91, 106, 127
embeddedness, 38, 76
emerging-market locations, 96, 99–100

face-to-face interaction, 18–19
factor market effects, 110–111
FDI externalities
  definition, 111
  inter-industry, 111
  intra-industry, 112
  pecuniary, 111
  non-pecuniary, 111
FDI motivations, 118
FDI spillover
  definition, 111
  antecedents, 113–121
  intra-industry, 46
  inter-industry, 46
  pipeline models, 118–119
flat world view, 4
Florida, R., 20, 43,

geographical systems of innovation, 91–94

headquarters, 35, 37, 45, 54–57, 63–65, 67–69, 72–74, 78, 106–110
heterarchical model, 57

Iammarino, S., 18, 20, 45, 87–91, 93, 95, 98
innovation studies literature, 79, 106, 127
internalization advantage, 45, 88

Jacobs, J., 46, 95

Index

knowledge creation, 80–83
knowledge protection, 83–85, see also *protection of firm-specific technology*
Kuemmerle, 36, 39, 40, 42

liability of foreignness, 89
liability of outsidership, 82
local buzz, 19, 96
local competition, 97–98
local firms absorptive capacity, 114, 116
local linkages, 76
location choices, 45–51, 102–104, 108, 117
locational hierarchy model, 48

MAR model, 97
Marshall, A., 46, 94–95
McCann, P., 18, 20, 87, 89–91, 93, 95, 98, 102
models of innovation in multinational firms, 12-15
Mudambi, R., 70–71, 74, 78, 87–89, 96, 99, 100, 102, 103
multilevel framework, 107–110, 123, 128, 130–132

oligopolistic deterrence, 102
organizational complexity, 52–54
ownership advantage, 10, 77, 87–88, 113, 118

Pearce, R.D., 24, 35, 41, 43
peripheral locations, 99
personal relationships, 96
physical attraction logic, 103–104
pipelines, 96
place and space, 89, 101, 127
Porter, M.E., 75, 94, 97–98,
processes of MNC innovation, 14–15
product cycle model, 12–13
product market effects, 110–111
proximity
    geographical, 16–21, 77–78, 94, 97, 125

    social, 19, 96, 125
    organizational, 19, 21
    institutional, 19
    cognitive, 19, 95, 97, 125
    external and internal, 21
protection of firm-specific technology, 36–37, 83–85, 106, 120–122

reciprocity, 85
related variety, 95
research and development
    adaptive, 11, 15, 38, 40–44, 51
    creative, 13–15, 40–43, 51
    in multinational firms, 10, 11
    internationalization, 10, 15–16
    trends, 22–33
R&D units
    generating station, 42
    home-base augmenting, 42, 46, 47
    home-base exploiting, 42, 46, 47
    international adaptor, 43
    international creator, 43
    listening posts, 42
    local adaptor, 43
    market-oriented units, 41
    market-seeking, 47
    multi-motive units, 42
    politically motivated units, 41–42
    production support, 41
    technology seeking, 47
relationships between R&D and foreign expansion, 10–11

Santangelo, G., 20, 99–100, 102
scale and scope economies in R&D, 36
sheer ignorance perspective, 56–57
spatial transaction costs, 18, 89
spiky world view, 49, 57
sub-national locations, 89–90, 91, 93, 108–110, 115, 129
subsidiary
    active and passive, 65–72
    center of excellence, 70
    competence-creating, 70–71

subsidiary – *Continued*
  competence-exploiting, 70–71
  entrepreneurship, 69
  power, 78–79
  roles, 65–66
  strategy, 65, 120, 122
  supply-driven forces, 39, 40–45

tacit knowledge, 16–20, 36–38, 77, 125
technological gap, 113–114
types of innovation-driven FDI, 40–45

Vernon, 12–13, 15–16

world mandate, 66